# A Primer on Multiple Intelligences

Matthew N. O. Sadiku • Sarhan M. Musa

# A Primer on Multiple Intelligences

Springer

Matthew N. O. Sadiku
Prairie View A&M University
Prairie View, TX, USA

Sarhan M. Musa
Prairie View A&M University
Prairie View, TX, USA

ISBN 978-3-030-77583-4 ISBN 978-3-030-77584-1 (eBook)
https://doi.org/10.1007/978-3-030-77584-1



This Springer imprint is published by the registered company Springer Nature Switzerland AG
The registered company address is: Gewerbestrasse 11, 6330 Cham, Switzerland

*To our children:*
*Motunrayo, Ann, and Joyce*
*Mahmoud, Ibrahim, and Khalid*

# Preface

The concept of intelligence is interesting and is discussed every day. It has been central to the field of psychology and remains hotly debated. It was first introduced by Francis Galton in 1885. Intelligence or cognitive development is a bio-psychological potential to process information that can be activated to solve problems. It is the capacity of the individual to act purposefully, to think rationally, and to deal effectively with his environment. The characteristic of intelligence is usually attributed to humans. Traditionally, intelligence is often regarded as a person's intellectual capacity; something one is born with and that cannot be changed. It is the ability to learn, to profit from experience, the ability to adapt to new situations, and the ability to solve problems. It was generally believed that intelligence was a single entity that was inherited. It is fixed and can be measured.

Historically, intelligence has been measured using the IQ test which is a general measure of one's cognitive function. Other views of intelligence have emerged in recent years. Today, an increasing number of researchers, believe that there exists a multiple of intelligences, quite independent of each other. People have a unique blend of intelligences. The concept of multiple intelligences represents an effort to reframe the traditional conception of intelligence.

Professor Howard Gardner at Harvard's Graduate School of Education argued that there are better or alternative ways to measure intelligence than standard IQ tests. He first proposed nine intelligences and insisted that all people are born with one or more intelligences. He believed human intelligence was multidimensional. This spectrum of intelligence indicates that people can be smart in a number of different ways. It implies that we are all intelligent in different ways, but there is always a primary, or more dominant, intelligence in each of us. Multiple intelligences theory states that everyone has all intelligences at varying degrees of proficiency, while most will experience more dominant intelligences that impact the way they learn and interact with others. Each type of intelligence presents a distinct component of our total competence. There is no clear consensus about how many intelligences there are. In this book, we consider 19 multiple intelligences. The intelligences are possessed by everyone and vary in degree of development within each individual.

The book is divided into three parts. Part I deals with human intelligence and consists of eight chapters. Chapter 1 presents Gardner's multiple intelligences as well as other intelligences. Chapter 2 deals with general intelligence (the most popular intelligence), which refers to a critical cognitive ability that has been well characterized by cognitive neuroscientists and psychologists. It is measured by psychological tests and is often stated in the form of intelligence quotient (IQ). Historically, general intelligence is regarded as the best individual predictor of achievement across all academic domains.

Chapter 3 covers naturalistic intelligence, which refers to a person's interest in the "natural" world of animals, plants, and the natural work around them. It is the ability to identify and classify patterns in natural environment. Chapter 4 deals with social intelligence, which is the ability to understand and manage men and women, boys and girls, and to act wisely in human relations or social situations. It is the ability to read nonverbal cues or make accurate social inferences. Chapter 5 is on spiritual intelligence, which implies awareness of our relationship to the divine. It is a unique ability to feel, understand, and act beyond selfish motives. It is the ability to behave with compassion and wisdom while maintaining peace regardless of the circumstances. Chapter 6 is concerned with emotional intelligence, which is a type of social intelligence that combines emotion and intelligence by viewing emotions as useful sources of information that help us to navigate our social environment. This is perhaps the most popular intelligence besides general intelligence. Chapter 7 is on interpersonal intelligence, which refers to the ability to relate well with others and manage relationships. It is the form of intelligence that governs your ability to relate to others and manage relationships. Chapter 8 dwells on intrapersonal intelligence (self smart), which is the capacity to explore one's inner world and feelings. It is essentially the ability to see realistically and honestly who we are and what we want. It is the type of intelligence we exercise when we try to learn more about ourselves. Chapter 9 discusses cultural intelligence, which is the capability to relate and work effectively across cultures. It is about understanding ourselves and understanding others. It is an ability that varies with individuals and explains differences in intercultural effectiveness.

Part II is basically on machine intelligence and consists of four chapters. Machines are performing better and better at analyzing data. Chapter 10 is on machine intelligence, which refers to the intelligence exhibited by computers. It results from the development of artificial intelligence and is often presented as artificial intelligence. The goal of machine intelligence is giving machines the ability to learn from the changing environment and making predictions. Chapter 11 is on artificial intelligence, which is the capacity to make computers learn from experience. It is a science with research activities in the areas of image processing, natural language processing, robotics, machine learning, etc. Although AI is a branch of computer science, there is hardly any field which is unaffected by this technology. Chapter 12 covers computational intelligence, which refers to recreating human-like intelligence in a computing machine. It consists of techniques that try to mimic human behavior, such as perception, reasoning, learning, evolution, and adaptation. Chapter 13 deals with digital intelligence, which refers to the combination of social, emotional, and cognitive abilities that enable individuals to adapt to the demands of life in the digital age. It is the use of generated data by digital technologies to improve the user experience. This is the kind of intelligence everybody ought to have in the twenty-first century.

Part III addresses other types of intelligences and consists of seven chapters. Chapter 14 is on business intelligence, which refers to the tools and systems that play a major role in the planning process of an organization. The tools represent business intelligence in customer support, customer profiling, market research, product profitability, and business analysis. Chapter 15 is on augmented intelligence, whose ultimate goal is using machines to enhance the work, expertise, and experience of a human user or worker. It is essentially artificial intelligence with a twist. Chapter 16 covers abstract intelligence, which is the ability to respond to words, numbers, letters, etc. It is the ability to carry on abstract thinking. It is a measure of one's ability to reason and understand complex concepts and assimilate new information beyond previous experience. It is independent of educational and cultural background. Chapter 17 addresses swarm intelligence, which is a complex discipline that seeks to understand and replicate both natural and artificial systems that are composed of several parts but effectively coordinate their activities through decentralized control. This type of intelligence is often displayed in nature through the habits of animals like bees, fish, birds, and ants. It enables individuals or entities to converge and create an organism that can do things that individuals cannot do by themselves. Chapter 18 deals with ambient intelligence, which is regarded as one of the emerging technologies that work with artificial intelligence techniques and wireless sensor networks to bring intelligence to our everyday electronic environments and makes those environments sensitive to humans. Its goal is to enhance the way people interact with their environment to promote safety and to enrich their lives. Chapter 19 discusses musical intelligence, which consists of sensitivity toward rhythm, melody, and musical instruments. It is the ability to enjoy, observe, differentiate, create, form, and express musical forms. Chapter 20 is on military intelligence, which is crucial and central to planning a victorious campaign. It provides guidance and direction to assist commanders in their decisions. The Appendix contains other types of intelligence that are less popular and not covered in this book.

This book provides researchers, students, and professionals a comprehensive introduction, applications, benefits, and challenges for 19 popular multiple intelligences. The authors were motivated to write this book partly due to the lack of a single source of reference on these popular multiple intelligences. Hence, the book will help a beginner to have an introductory knowledge about these intelligences. The main objective of the authors is to provide a concise treatment that is easily digestible for each intelligence. It is hoped that the book will be useful to psychologists, computer scientists, and the general public.

Prairie View, TX, USA

Matthew N. O. Sadiku
Sarhan M. Musa

# Contents

# About the Authors

**Matthew N. O. Sadiku** received his B.Sc. degree in 1978 from Ahmadu Bello University, Zaria, Nigeria, and his M.Sc. and Ph.D. degrees from Tennessee Technological University, Cookeville, TN, in 1982 and 1984 respectively. From 1984 to 1988, he was an assistant professor at Florida Atlantic University, Boca Raton, FL, where he did graduate work in computer science. From 1988 to 2000, he was at Temple University, Philadelphia, PA, where he became a full professor. From 2000 to 2002, he was with Lucent/Avaya, Holmdel, NJ, as a system engineer and with Boeing Satellite Systems, Los Angeles, CA, as a senior scientist. He is presently a professor emeritus of electrical and computer engineering at Prairie View A&M University, Prairie View, TX.

He is the author of over 900 professional papers and over 90 books including *Elements of Electromagnetics* (Oxford University Press, seventh ed., 2018), *Fundamentals of Electric Circuits* (McGraw-Hill, seventh ed., 2021, with C. Alexander), *Computational Electromagnetics with MATLAB* (CRC Press, fourth ed., 2019), *Principles of Modern Communication Systems* (Cambridge University Press, 2017, with S. O. Agbo), and *Emerging Internet-Based Technologies* (CRC Press, 2019). In addition to the engineering books, he has written Christian books including *Secrets of Successful Marriages*, *How to Discover God's Will for Your Life*, and commentaries on all the books of the New Testament Bible. Some of his books have been translated into French, Korean, Chinese (and Chinese Long Form in Taiwan), Italian, Portuguese, and Spanish.He was the recipient of the 2000 McGraw-Hill/Jacob Millman Award for outstanding contributions in the field of electrical engineering. He was also the recipient of Regents Professor award for 2012–2013 by the Texas A&M University System. He is a registered professional engineer and a fellow of the Institute of Electrical and Electronics Engineers (IEEE) "for contributions to computational electromagnetics and engineering education." He was the IEEE Region 2 Student Activities Committee Chairman. He was an associate editor for *IEEE Transactions on Education*. He is also a member of Association for Computing Machinery (ACM) and American Society of Engineering Education (ASEE). His current research interests are in the areas of computational electromagnetics, computer networks, and engineering education. His works can be found in his autobiography, *My Life and Work* (Trafford Publishing, 2017) or his website: www.matthew-sadiku.com. He currently resides in Hockley, Texas. He can be reached via email at sadiku@ieee.org.

**Sarhan M. Musa** is a professor in Electrical and Computer Engineering Department at Prairie View A&M University. He holds a Ph.D. in Electrical Engineering from the City University of New York. He is the founder and director of Prairie View Networking Academy (PVNA), Texas. He is LTD Sprint and Boeing Welliver Fellow. Professor Musa is internationally known through his research, scholarly work, and his published books. He has had a number of invited talks at international conferences. He has received a number of prestigious national and university awards and research grants. He is a senior member of the IEEE and has also served as the member of technical program committee and steering committee for a number of major journals and conferences. Professor Musa has written more than a dozen books on various areas in Electrical and Computer Engineering. His current research interests cover many topics in artificial intelligence/ML, data analytics, Internet of things, wireless network, data center protocols, and computational methods.

# Chapter 1
# Introduction

## Contents



Reason is intelligence taking exercise. Imagination is intelligence with an erection.—
Victor Hugo

## 1.1 Introduction

The concept of intelligence has been central to the field of psychology and remains hotly debated. The characteristic of intelligence is usually attributed to humans. Traditionally, intelligence is often regarded as a person's intellectual capacity; something one is born with and that cannot be changed. It is often regarded as the ability to solve problems. It was generally believed that intelligence was a single entity that was inherited. It is fixed and can be measured. For a human potential to be identified as an intelligence, it must satisfy the following eight criteria [1]:

1. It must be rooted in the brain, so that an injury to the brain could rob a person of that specific potential. Different intelligences are located in different parts of the brain.
2. It must be rooted in our evolutionary history such that our early ancestors exhibited that potential.
3. There has been an identifiable core operation or set of operations associated with that potential.


M. N. O. Sadiku, S. M. Musa, *A Primer on Multiple Intelligences*,
https://doi.org/10.1007/978-3-030-77584-1_1

4. It must be susceptible to being encoded in symbol (e.g., mathematical symbols).
5. It must possess a distinctive developmental path to become expert in that ability.
6. It is exemplified by the existence of idiot savants, prodigies, and other exceptional people.
7. There is evidence from experimental psychology that the ability is distinct from other abilities.
8. It is supported by psychometric findings.

Historically, intelligence has been measured using the IQ test which is a general measure of one's cognitive function. Other views of intelligence have emerged in recent years. Today, an increasing number of researchers, including Howard Gardner, believe that there exists a multiple of intelligences, quite independent of each other. People have a unique blend of intelligences [2]. The concept of multiple intelligences represents an effort to reframe the traditional conception of intelligence.

Professor Howard Gardner at Harvard's Graduate School of Education argued that there are better or alternative ways to measure intelligence than standard IQ tests. He believed human intelligence was multidimensional. He was first to propose the theory of multiple intelligences in his 1983 famous book [3]. He shook the educational world by what he called multiple intelligences. These are: verbal/linguistic, mathematical/logical, spatial, musical, kinesthetic, interpersonal, and intrapersonal. In 2006 (23 years later), Gardner added an eighth branch to his model in his another book [4]. This eighth branch is known as naturalistic intelligence. Gardner has suggested the possible addition of a ninth known as existentialist intelligence. This spectrum of intelligence indicates that people can be smart in a number of different ways. It implies that we are all intelligent in different ways, but there is always a primary, or more dominant, intelligence in each of us.

This chapter presents Gardner's multiple intelligences as well as other intelligences such as social, emotional, spiritual, military, and artificial intelligences. It discusses some applications of multiple intelligences. It presents the applications of multiple intelligence at the global level. It highlights the benefits and challenges of multiple intelligences. This chapter also includes some quotations or intellerge. The last section concludes with comments.

## 1.2 Gardner's Intelligence

Multiple intelligence (MI) theory pluralizes the traditional concept of intelligence. As mentioned earlier, Howard Gardner developed the theory of multiple intelligences. He listed the following forms of intelligence [4–6]:

1. *Musical intelligence*: The capacity to think and feel in terms of sound and rhythm. Individuals with this intelligence, such as Michael Jackson and Ed Sheeran, can recognize and create musical pitch, rhythm, timbre, and tone.
2. *Spatial Intelligence:* The ability to see and represent the world using concepts such as shape, color, and form. Individuals with this intelligence, such as Frank

Lloyd Wright and Amelia Earhart, can recognize and manipulate fine-grained spatial images.
3. *Mathematical/Logical reasoning*: The ability for working with numbers, logical concepts, and abstract analysis. People with this intelligence, such as Albert Einstein and Bill Gates, can readily develop equations, perform calculations, and solve abstract problems.
4. *Interpersonal Intelligence*: The ability to understand and interact with other people effectively. Individuals with this intelligence, such as Mahatma Gandhi and Mother Teresa, can recognize and understand other people's moods, desires, motivations, and intentions.
5. *Intrapersonal Intelligence:* A person's innate ability to understand his or her inner world, a world from which many people are entirely disconnected. People with this intelligence, such as Aristotle and Maya Angelou, can recognize and understand their own moods, desires, motivations, and intentions.
6. *Bodily-Kinesthetic Intelligence*: The ability for motor skills, such as dancers, athletes, sculptors, or even surgeons. People with this intelligence, such as Michael Jordan and Simone Biles, can use one's own body to create products, perform skills, or solve problems through mind–body union.
7. *Linguistic Intelligence:* The ability to express oneself with such clarity that others can comprehend and relate to them. People with this intelligence, such as William Shakespeare and Oprah Winfrey, can analyze information and easily create products such as speeches, books, and memos.
8. *Naturalistic Intelligence*: The ability to perceive and classify the intricate patterns in the world around them. Individuals with this type of intelligence, such as Charles Darwin and Jane Goodall, can identify and distinguish among different types of plants, animals, and weather formations that are found in the natural world.
9. *Existential Intelligence*: The sensitivity and capacity to tackle deep questions about human existence, such as the meaning of life, why we die, and how we got here. Existential intelligence includes aesthetic, philosophy, and religion and emphasizes the classical values of beauty, truth, and goodness.

Multiple intelligences theory states that everyone has all nine intelligences at varying degrees of proficiency, while most will experience more dominant intelligences that impact the way they learn and interact with others. Each type of intelligence presents a distinct component of our total competence. We can think of them as separate for purposes of discussion, but actually they are intimately interwoven. That implies that a child may be highly proficient in one intelligence and struggle with another. All nine intelligences are needed to live a balanced life. They are amoral in that they can be put to constructive or destructive use. Figure 1.1 illustrates the nine multiple intelligences.

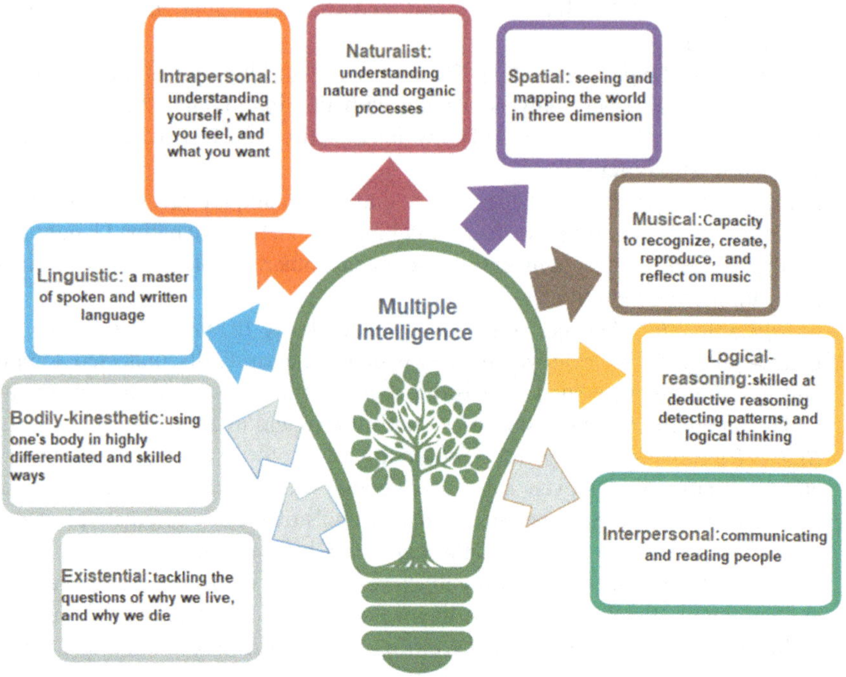


**Fig. 1.1** Gardner's theory of nine multiple intelligences

## 1.3 Other Types of Intelligences

Evidently, Gardner's list is not exhaustive. There are other types of intelligence such as social intelligence, spiritual intelligence, emotional intelligence, computational intelligence, machine intelligence, artificial intelligence, swarm intelligence, ambient intelligence, digital intelligence, military intelligence, and the list continues. These intelligences are relatively independent. Every intelligence is unique and important in its right. In this section, we will focus on some of these additional intelligences.

- *Emotional Intelligence:* This refers to the ability to identify and manage one's own emotions, as well as the emotions of others. It is a set of abilities related to the understanding, use, and management of emotion as it relates to one's self and others. People have different personalities, traits, and values, and different ways of expressing their emotions. We unconsciously use emotional intelligence in our daily lives [7].
- *Spiritual Intelligence:* This is concerned with integrating the inner life of mind and spirit with the outer life of work in the world. It is essential for both personal and professional development. The term implies awareness of our relationship to the divine. It is a unique ability to feel, understand, and act beyond selfish

motives. It can be expressed as love, compassion, modesty, wisdom, and service [8].

- *Social Intelligence:* This refers to the ability to successfully build relationships and navigate social environments. It is about figuring out the best way to get along with others. It is the ability to adequately understand and evaluate one's own behavior and the behavior of others. It is the ability to get along well with others and win their cooperation. It is the key to life and career success [9].
- *Artificial Intelligence* (AI): This is the cognitive science that deals with intelligent machines which are able to perform tasks heretofore only performed by human beings. It is mainly concerned with applying computers to tasks that require knowledge, perception, reasoning, understanding, and cognitive abilities. AI is potentially the algorithmic study of processes in every field of study. The main objective of AI is to teach the machines to think intelligently like humans do [10].
- *Swarm Intelligence:* This is the emergent collective intelligence of groups of simple agents. The concept of swarm intelligence is an emerging domain which belongs to the field of artificial intelligence (AI), as many of its pursuits can be linked to machine learning. It belongs to the emerging field of bio-inspired soft computing, which is closely related to AI. It is inspired from the biological entities such as birds, fish, ants, wasps, termites, and bees [11].
- *Computational Intelligence.* This refers to recreating human-like intelligence in a computing machine. It consists of a set of computing systems with the ability to learn and deal with new situations such that the systems are perceived to have some attributes of intelligence. It is efficient in solving real-world problems which require reasoning and decision-making. It produces more robust, simpler, and tractable solutions than the traditional techniques. The goal of computational intelligence is to recreate human-like intelligence in a human-made machine [12].
- *Digital Intelligence:* This refers to an all-encompassing set of technical, cognitive, social, and emotional competencies that enable individuals to face the challenges of digital life. It belongs to "meta-intelligence"—one that is composed of many constituent intelligences. Digital intelligence addresses the what, why, where, when, who, how, and how much of digital technology to improve our operational efficiency and performance. It is about putting the data to use immediately. It is the intersection of big data with big science.
- *Military Intelligence:* This is a military branch that uses information collection and analysis approaches to provide guidance and direction to assist commanders in their decisions. As an academic field, military intelligence is multidisciplinary area that combines language, political theory, economics, sociology, and psychology. Military intelligence includes information on other nations' military forces, plans, and operations which is collected through a variety of means. The term intelligence is often used for an agency that gathers such information.

Other forms of intelligence include mobile intelligence, industrial intelligence, abstract intelligence, concrete intelligence, cultural intelligence, machine intelligence, human intelligence, business intelligence, augmented intelligence, Internet intelligence, public intelligence, competitive intelligence, and the list continues.

## 1.4 Applications of Multiple Intelligences

Multiple intelligence theory aims to better represent the diversity in human intelligence. It has significant implications for education, use of technology, architecture, and leadership.

- *Education:* The theory of multiple intelligences was born in the field of education (or educational psychology). It states that each person has different ways of learning and different intelligences they use in their daily lives. Educators have positively responded and embraced Gardner's theory. The multiple intelligences theory can be used for curriculum development, designing instruction, selection of course activities, and assessment strategies. The theory proposes a major transformation in the way our schools operate. It suggests that teachers be trained to teach in a wide variety of ways using music, cooperative learning, art activities, role play, multimedia, field trips, inner reflection, etc. As shown in Fig. 1.2, understanding multiple intelligences helps teachers to understand children better. Gone are the days when educators could rely on an IQ test to measure a student's intelligence. Since students learn in different ways, teachers who are committed to excellence continually seek different ways to improve teaching and learning. As shown in Fig. 1.3, the bridge between existing psychological models and education is spanned by multiple intelligences. Theory of multiple intelligences suggests several ways in which the material might be presented to facilitate effective learning [13]:

- □ Words (linguistic intelligence).
- □ Numbers or logic (logical-mathematical intelligence).
- □ Pictures (spatial intelligence).

**Fig. 1.2** Understanding multiple intelligences in children

**Fig. 1.3** The bridge between existing psychological models and education is spanned by multiple intelligences

- Music (musical intelligence).
- Self-reflection (intrapersonal intelligence).
- A physical experience (bodily-kinesthetic intelligence).
- A social experience (interpersonal intelligence).
- An experience in the natural world (naturalist intelligence).

- *Use of Technology:* MI and technology blend in the modern, rapidly changing environment of education. Thoughtful and purposeful use of technology has a great impact on student achievement. A curriculum that incorporates technology and MI theory supplements students' strengths and expands their possibilities [14]. Effective instruction directly affects students' acquisition of the twenty-first century skills necessary for them to compete in increasingly competitive workplace. With the advent of new technologies, brain-based researchers have started to look inside the brain to see activity and movement using advanced equipment. These technological tools have allowed researchers to better understand which parts of the brain are dormant and gain new insight into how the brain develops, learns, and processes information [15]. For example, electronic storybooks are becoming common in education as one way to teach students. Multimedia and the Internet incorporate all the elements of skill work.
- *Architecture*: By nature and tradition, architectural education holds potential for integration and application of learning and instilling intellectual rigor. The complexity of architectural design tasks requires individuals who possess a wide

range of abilities, such as spatial visualization, logical thinking, problem-solving skills, linguistic ability, communication skills, etc. The theory of "multiple intelligences" has been introduced in architecture design [16].

- *Leadership:* Leaders induce followers to act for certain goals that represent the values and the motivations. It is critical to the success of leaders that the relationship between intelligence and leadership effectiveness is understood. There is a relationship between a leader's intelligence and their leadership behaviors [17]. Multiple intelligences become the tools that help the leader adjust and tweak his role. Transformational leader uses his/her multiple intelligences to raise followers' consciousness about goals and values.
- *Business:* Globalization has led to the integration of goods, services, and culture among the nations of the world. This implies that business schools must equip their graduates with the skills to operate across national boundaries. Business educators must understand that intelligences can be nurtured and developed, and that students' strengths can change with time. The value of the MI theory lies in its ability to cause educators to reflect upon teaching and learning, and to ultimately become better educators [18].

MI theory can also be applied in dance education, music education, legal education sports, counseling, and fine art.

## 1.5 Global Multiple Intelligences

Gardner's theory of multiple intelligences has become a rallying point for educators throughout the world. It has had enormous influence on educational thinking and practice around the world. Culture dictates which of the intelligences are highly valued in a given society. Cultural practices, beliefs, attitudes, rituals, customs, and values all influence the definition and characteristics of the concept of intelligence. We now consider how multiple intelligences are applied in different nation.

- *United States:* The multiple intelligences (MI) theory has had a profound impact on thinking and practice in education in the United States. Many schools in the USA have sought to structure curricula according to the multiple intelligences, offer teachers an exciting opportunity to appeal to the different personalities and learning styles represented in the classroom. MI has made the curriculum and instruction far more inclusive of various learning styles. Learning about MI theory changed the thinking of teachers and students. Teachers now think of ability in an entirely different way, students embrace the concept because it celebrates their diverse talents [19].
- *China:* In the development of gifted education, educators in Hong Kong have broaden the concept of gifted to include different aspects of giftedness. Hong Kong educators regard the development of multiple intelligence. In gifted education, multiple intelligences theory has implications for identification, learning, teaching, assessment, and evaluation [20].

- *Australia*: MI theory has made a strong influence in a wide range of educational settings in Australia. It has transformed education in Australia, especially in elementary and gifted education. The theory also helped teachers shift assessment from narrowly based process to a broader understanding. MI schools have transformed their relationship with parents [21].
- *Qatar:* This is a small peninsula with its only land-link through Saudi Arabia. In 2008, the world witnessed the lack of expatriate commitment with the mass exodus of expatriates in Dubai. Companies seek global expatriate workers for general skills such as accounting, management, business development, marketing, and construction. These expatriates working in Qatar are employees working away from his/her nation of nationality. Intelligence is understood as the decision-making process, where personality is the behavior displayed [22].
- *South Africa:* To meet the demands of a highly technological and globally competitive society, all learners in South Africa must obtain skills in mathematics. Education in South Africa now faces the challenge of how to improve the mathematics performance of students. An approach to deal with this challenge is to apply MI theory in the classroom, with learners coming from diverse backgrounds. Teachers can use MI theory in the classroom to change their teaching and learning strategies, but this is no simple task [23].
- *Irish:* The engagement of the Irish education system with the MI has been phenomenal. The application of MI theory to enhancing teaching, learning, and assessment in schools, colleges, and universities is mediated through individual subject syllabuses with prioritizing the linguistic and logical-mathematical intelligences at the expense of other intelligences. The insights of the Irish MI project suggest that the ideas educators hold about intelligence have more power than might be imagined in how young people are valued and educated [24].

## 1.6 Benefits

Multiple intelligences challenge the traditional view of intelligence. The standard psychological view of intellect states that there is a single intelligence, adequately measured by IQ, while multiple intelligences (MI) theory claims that human beings have a number of relatively discrete intellectual capacities.

Multiple intelligences guide the way we learn and process information. The traditional mode of teaching has not been successful for all students as is evidenced by the dropout rate of 50% in high schools in the United States. In the twenty-first century, a combination of new technology, new knowledge, and rapid access to that knowledge is the key to individualizing instruction, but this will require new teaching methods to make the best use of available resources [18].

Gardner's theory of multiple intelligences enjoys considerable popularity among educators who have incorporated multiple intelligences in their teaching. It has been suggested as alternative to more traditional psychometric tests. It has helped educators to question their work, reflect on their practice, and to encourage them to use

MI as a means to foster high-quality work. Multiple intelligences confirm what educators already experience: students think and learn in many different ways. Nine kinds of intelligence would allow nine ways to teach, instead of one.

## 1.7 Challenges

Gardner's theory of multiple intelligences has not been widely accepted within academic psychology. It has come under criticism from both psychologists and educators for lacking supporting empirical research. Most of the criticism and resistance come from cognitive psychologists, psychometricians, and from those removed from the classroom, such as journalists and researchers. In spite of the impact of the theory of multiple intelligences, there are still thousands of schools that follow the traditional way of teaching. The challenge is to get this information out to many more parents, teachers, and school administrators.

If someone claims that they teach to the multiple intelligences, one should ask how it is done, what instructional methods are followed, what assessment techniques are used [25]. The multiple intelligences seem to be limited to educational use and have not been widely applied elsewhere.

## 1.8 Quotes on Intelligence

> Reason is intelligence taking exercise. Imagination is intelligence with an erection.—Victor Hugo

> I am so clever that sometimes I don't understand a single word of what I am saying.—Oscar Wilde

> You might be poor, your shoes might be broken, but your mind is a palace.—Frank McCourt

> Whatever the cost of our libraries, the price is cheap compared to that of an ignorant nation.—Walter Cronkite

> The measure of intelligence is the ability to change.—Albert Einstein

> A good head and good heart are always a formidable combination. But when you add to that a literate tongue or pen, then you have something very special.—Nelson Mandela

> It takes something more than intelligence to act intelligently.—Fyodor Dostoyevsky

> It is one thing to be clever and another to be wise.—George R.R. Martin

> The ability to observe without evaluating is the highest form of intelligence.—J. Krishnamurti

Intelligence without ambition is a bird without wings.—Walter H. Cottingham

Education is no substitute for intelligence.—Frank Herbert

Crazy people are considered mad by the rest of the society only because their intelligence isn't understood.—Wei Hui

Human intelligence is richer and more dynamic than we have been led to believe by formal academic education.—Ken Robinson

Every machine has artificial intelligence. And the more advanced a machine gets, the more advanced artificial intelligence gets as well. But, a machine cannot feel what it is doing. It only follows instructions - our instructions - instructions of the humans. So, artificial intelligence will not destroy the world. Our irresponsibility will destroy the world.—Abhijit Naskar

Any fool can make things bigger, more complex, and more violent. It takes a touch of genius-and a lot of courage-to move in the opposite direction.—Albert Einstein

Creating something from nothing is ingenious. Creating something from something is smart. To rule your world, be ingeniously smart.—Martin Uzochukwu Ugwu

A scientist is proud of his intelligence, an artist is proud of his imagination.—Amit Kalantri

Hard work eventually solves all problems, but to solve problems quickly you need intelligence.—Amit Kalantri

Nature has distributed intelligence unequally, but imagination equally.—Amit Kalantri

## 1.9 Conclusion

The intelligences are possessed by everyone and vary in degree of development within each individual. Multiple intelligences refer to the concept introduced by Howard Gardner that challenges the traditional view of intelligence and explains the presence of nine different multiple intelligences in people. Howard Gardner is one of those researchers who have recently questioned the idea that intelligence is a single entity. Although some criticize and resist Howard Gardner's notion of multiple intelligences, it still has helped a significant number of educators to question their work and reflect on their practice [26].

While the concept of multiple intelligences still remains a theory, various ways of assessing the intelligences have been developed. These include The Rogers Indicator of Multiple Intelligences (RIMI), Tree of Intelligences (TOI), the Teele Inventory of Multiple Intelligences (TIMI), the Multiple Intelligence Developmental Assessment Scales for Children (MIDAS-for-KIDS), and the DISCOVER assessment [27].

Without doubt, MI is no fad, but a proven approach to education for the twenty-first century. For more information about multiple intelligences, please visit Howard Gardner's official website of Multiple Intelligences Theory at www.multipleintelligencesoasis.org. One may also consult the books in [3, 4, 28–59] and the following related journal: *Journal of Intelligence.*

## References


1. S.J. Denig, Multiple intelligences and learning styles: two complementary dimensions. Teach. Coll. Rec. **106**(1), 96–111 (2004)
2. M.N.O. Sadiku, S.M. Musa, A. Ajayi-Majebi, A primer on multiple intelligence. Int. J. Trend Res. Develop. **7**(4) (2020)
3. H.E. Gardner, *Frames of Mind: The Theory of Multiple Intelligence* (Basic Book, New York, 1983)
4. H.E. Gardner, *Multiple Intelligences: New Horizons in Theory and Practice* (Basic Books, 2006)
5. D. Gutierrez, K. Perri, A. Quackenbush, Exploring the multiple intelligences of community college students enrolled in online courses. J. Coll. Teach. Learn. **3**(11), 85–90 (2006)
6. M. Marenus, Gardner's theory of multiple intelligences. Simply Psychol. (2020), https://www.simplypsychology.org/multiple-intelligences.html
7. M.N.O. Sadiku, O.D. Olaleye, S.M. Musa, Emotional intelligence: a primer. Int. J. Trend Res. Develop. **7**(1), 109–112 (2020)
8. M.N.O. Sadiku, A. Omotoso, S.M. Musa, Toward better understanding of spiritual intelligence. Int. J. Trend Scient. Res. Develop. **4**(4), 1113–1115 (2020)
9. M.N.O. Sadiku, S. Alam, S.M. Musa, Social intelligence: a Primer. Int. J. Res. Granthaal. **7**(9), 213–217 (2019)
10. M.N.O. Sadiku, S.M. Musa, O.M. Musa, Artificial intelligence in the manufacturing industry. Int. J. Adv. Scient. Res. Eng. **5**(6), 108–110 (2019)
11. M.N.O. Sadiku, M. Tembely, S.M. Musa, Swarm intelligence. Int. J. Adv. Res. Comp. Sci. Soft. Eng. **8**(5), 100–102 (2018)
12. M.N.O. Sadiku, J. Foreman, S.M. Musa, Computational intelligence. Eur. Sci. J. **14**(21), 56–60 (2018)
13. E. Herndon, What are multiple intelligences and how do they affect learning? (2018), https://www.cornerstone.edu/blogs/lifelong-learning-matters/post/what-are-multiple-intelligences-and-how-do-they-affect-learning
14. I.J. McCoog, Integrated instruction: multiple intelligences and technology. Clear. House J. Educ. Strateg. Issues Ideas **81**(1), 25–28 (2007)
15. C.D. Mullican, Multiple intelligences in the text: examining the presence of multiple intelligences tasks in the annotated teacher's editions of four high school United States history textbooks, Doctoral Dissertation, University of South Florida, 2012
16. M.S. D'souza, The complexity of architectural design tasks requires individuals who possess a wide array of abilities - for example, spatial visualization, logical thinking, problem solving skills, linguistic ability, communication skills and so on, Doctorol Dissertation, The University of Wisconsin-Milwaukee, 2006
17. J.B. Bullock, Intelligence and leadership: an investigation of multiple intelligences as antecedents to transactional and transformational leadership behaviors, Doctoral Dissertation, Marian University, 2008
18. A. Vincent, D. Ross, A. Williams, Using the multiple intelligences theory in international business. J. Teach. Int. Bus. **14**(1), 45–63 (2002)

19. G. Mettetal, C. Jordan, S. Harper, Attitudes toward a multiple intelligences curriculum. J. Educ. Res. **91**(2), 115–122 (1997)
20. D.W. Chan, Perceived multiple intelligences among male and female Chinese gifted in Hong Kong: the structure of the multiple intelligences profile. Gift. Child Q. **50**(4), 325–356 (2006)
21. W. Vialle, In Australia: multiple intelligences in multiple settings. Educ. Leadersh. **55**(1), 65–69 (1997)
22. D. Smit, Expatriate tenure and multiple intelligences in Qatar, Doctoral Dissertation, University of Phoenix, 2014
23. E. Gouws, A.M. Dicker, Teaching mathematics that addresses learners' multiple intelligences. Africa Educ. Rev. **8**(3), 568–587 (2011)
24. J. Hanafin, Multiple intelligences theory, action research, and teacher professional development: the Irish MI Project. Aust. J. Teach. Educ. **39**(4), 126–141 (2014)
25. S.R. Koff, Why the multiple intelligences? J. Dance Educ. **3**(1), 5–6 (2003)
26. M.K. Smith, *Howard Gardner and Multiple Intelligences* (Encyclopedia of Pedagogy and Informal Education, 2008)
27. S.D. McMahon, D.S. Rose, M. Parks, Multiple intelligences and reading achievement: an examination of the Teele inventory of multiple intelligences. J. Exp. Educ. **73**(1), 41–52 (2044) https://www.infed.org/mobi/howard-gardner-multiple-intelligences-and-education
28. H. Gardner, *The Shattered Mind* (Knopf, New York, 1975)
29. H. Gardner, *Changing Minds. The Art and Science of Changing our Own and Other People's Minds* (Harvard Business School Press, Boston, MA, 2006)
30. W. McKenzie, *Multiple Intelligences and Instructional Technology*, 2nd edn. (International Society for Technology in Education, 2005)
31. J.L. Kincheloe, *Multiple Intelligences Reconsidered* (International Academic Publishers, 2004)
32. T. Armstrong, *The Multiple Intelligences of Reading and Writing: Making the Words Come Alive* (ASCD, 2003)
33. R.J. Sternberg, *Successful intelligence* (Simon & Schuster, New York, 1996)
34. T. Armstrong, *Multiple Intelligences in the Classroom*, 4th edn. (ASCD, Alexander, VA, 2017)
35. L. Kagan, S. Kagan, *Multiple Intelligences: The Complete MI Book* (Kagan Cooperative Learning, 1998)
36. R.E. Riggio, S.E. Murphy, F.J. Pirozzolo, *Multiple Intelligences and Leadership* (Taylor & Francis, 2001)
37. W. McKenzie, *Multiple Intelligences and Instructional Technology* (ISTE, 2005)
38. K. Andriotis, *Multiple Intelligences and Curriculum Development* (Tate Publishing & Enterprise, 2012)
39. H.F. Silver, R.W. Strong, M.J. Perini, *So Each May Learn: Integrating Learning Styles and Multiple Intelligences* (ASCD, 2000)
40. J.Q. Chen, S. Moran, H. Gardner, *Multiple Intelligences Around the World* (Wiley, 2009)
41. R.J. Fogarty, J. Stoehr, *Integrating Curricula with Multiple Intelligences: Teams, Themes & Threads*, 2nd edn. (Corwin Press, Thousand Oaks, CA, 2008)
42. T.R. Hoerr, N.C. Rolheiser-Bennett, *Becoming a Multiple Intelligences School* (ASCD (Association for Supervision and Curriculum Development), Alexander, VA, 2000)
43. J.L. Kincheloe, *Multiple Intelligences Reconsidered* (Peter Lang, New York, 2004)
44. L. Campbell, B. Campbell, *Multiple Intelligences and Student Achievement: Success Stories from Six Schools* (ASCD, Alexander, VA, 1999)
45. S. Baum, J. Viens, B. Slatin, *Multiple Intelligences in the Elementary Classroom: A Teacher's Toolkit* (Teachers College Press, New York, 2005)
46. J. Viens, S. Kallenbach, *Multiple Intelligences and Adult Literacy: A Sourcebook for Practitioners* (Teachers College Press, New York, 2004)
47. M. Kornhaber, E. Fierros, S. Veenema, *Multiple Intelligences: Best Ideas from Research and Practice* (Pearson, 2003)
48. K. Davis et al., *The Theory of Multiple Intelligences: Cambridge Handbook of Intelligence* (Cambridge University Press, New York, 2011)

49. T. Armstrong, *7 (Seven) Kinds of Smart: Identifying and Developing Your Multiple Intelligences*, 2nd edn. (Plume, 1999)
50. T. Armstrong, *You're Smarter Than You Think: A Kid's Guide to Multiple Intelligences*, 2nd edn. (Free Spirit Publishing, 2014)
51. J.A. Bellanca, *200+ Active Learning Strategies and Projects for Engaging Students' Multiple Intelligences*, 2nd edn. (Corwin, 2008)
52. H. Gardner, *Intelligence Reframed; Multiple Intelligences for the 21st Century* (Basic Books, 1999)
53. J. Martin, *Profiting from Multiple Intelligences in the Workplace* (Gower Publishing, 2001)
54. J. Martin, *Multiple Intelligences; a Trainer's Resource of 35 Activities* (Gower Publishing, 2001)
55. K. Tirri, P. Nokelainen, *Measuring Multiple Intelligences and Moral Sensitivities in Education* (Sense Publishers, 2011)
56. R.A. Hirsh, *Early Childhood Curriculum; Incorporating Multiple Intelligences, Developmentally Appropriate Practice, and Play* (Allyn & Bacon, 2004)
57. I.H. Buchen, *Futures Thinking, Learning, and Leading; Applying Multiple Intelligences to Success and Innovation* (Rowman & Littlefield Education, 2006)
58. B. Campbell, *Handbook of Differentiated Instruction Using the Multiple Intelligences; Lesson Plans and More* (Allyn & Bacon, 2008)
59. L. Campbell, B. Campbell, D. Dickinson, *Teaching and Learning Through Multiple Intelligences*, 3rd edn. (Allyn and Bacon, 2003)

# Part I
# Human Intelligence

# Chapter 2
# General Intelligence

**Contents**



> Everybody is a genius, but if you judge a fish by its ability to climb a tree, it will live its whole life believing that it is stupid.
>
> —Albert Einstein

## 2.1 Introduction

Individuals differ in their ability to reason, solve problems, think abstractly, plan, and learn. This general ability is known as intelligence. No topic in psychology has provoked more intense public controversy than the concept of intelligence. Humans have been interested in intelligence since our earliest written history. Intelligence can be regarded as a general mental ability that involves being able to reason, plan, solve problems, think abstractly, comprehend complex ideas, learn quickly, and learn from experience. It is a complex characteristic of cognition. The standard view in the field of human intelligence has been that there is a "general intelligence" (GI) that permeates all human cognitive activity. Intelligence is studied in humans, non-human animals, plants, and machines. High general intelligence (IQ) has evolved with convergent evolution in nonhuman animals such as capuchins, baboons, macaques, and apes.



M. N. O. Sadiku, S. M. Musa, *A Primer on Multiple Intelligences*, https://doi.org/10.1007/978-3-030-77584-1_2

There are consistent individual differences in human intelligence, attributable to a single "general intelligence" factor, *g*. Although intelligence is often studied in humans, it has been observed in animals and plants. In humans, performance on diverse tests of cognitive ability shows positive correlations with a single factor, termed general intelligence or "*g*" [1].

General intelligence, also known as *g* factor, refers to a critical cognitive ability that has been well characterized by cognitive neuroscientists and psychologists. It is one of the most pervasive concepts in psychology. It is measured by psychological tests and is often stated in the form of intelligence quotient (IQ). Historically, general intelligence is regarded as the best individual predictor of achievement across all academic domains.

Tests that are g-loaded are among the best predictors of success in school and workplace. Higher general intelligence is often associated with educational and occupational successes [2].

This paper provides an introduction to general intelligence. It begins by describing what general intelligence is all about. It discusses three aspects of intelligence and features of general intelligence. It provides measures of the intelligence quotient. It covers some applications of general intelligence and global general intelligence. It highlights the benefits and challenges of general intelligence. The last chapter concludes with comments.

## 2.2 What Is General Intelligence?

Intelligence refers to the ability to think, to learn from experience, to solve problems, and to adapt to new situations. It can be quantified across species and is heritable. Humans share with other species a core set of genetically specified constraints on intelligence. The traditional view in the field of human intelligence has been that there is a "general intelligence" that permeates all human cognitive activity. General intelligence (GI) accounts for the overall differences in intelligence among people. It is correlated with gender (men scored higher than women), age (younger scored higher elderly), and brain size (larger brains scored higher). This largely explains male dominance in most cultures and the idea that smarter people have bigger brains. Intelligence can have different meanings and values in different cultures. What is considered "intelligent" in one culture may be different from those considered intelligent in another.

In the 1900s, scientists reduced intelligence down to a single digit: the g factor. The idea of IQ is based on a "g factor" or "generalized intelligence." Social and economic deprivation can negatively affect IQ. Intelligence is improved by education.

General intelligence is a cognitive process in that it gives humans the cognitive abilities to learn, understand, and reason. As illustrated in Fig. 2.1, the various cognitive factors that make up general intelligence include [3]:

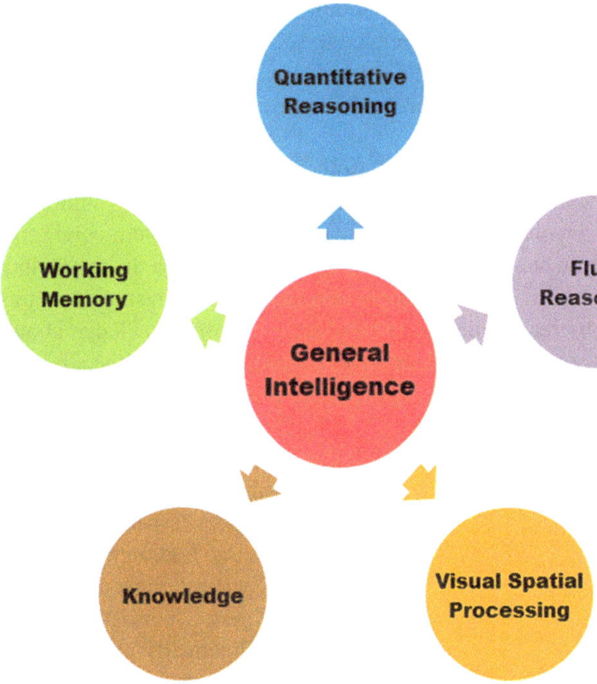


**Fig. 2.1** Various cognitive factors that make up general intelligence

- *Quantitative reasoning:* the test that involves capacity to solve numerical problems
- *Fluid reasoning:* flexible thinking to solve problems
- *Visual-spatial processing:* ability to put together puzzles and copying complex shapes
- *Knowledge:* a person's knowledge about vast array of topics
- *Working memory:* capacity of the short-term memory, such as repeating a list

## 2.3 Three Aspects of Intelligence

Several theories have been developed to explain what intelligence is and how it works. One popular theory is Howard Gardner's theory of multiple intelligences, discussed in the previous chapter. Another popular theory is Sternberg's triarchic theory of intelligence. According to British psychologist Robert J. Sternberg, intelligence has three aspects: analytical, creative, and practical [4, 5]:

- *Analytical Intelligence*, also referred to as componential intelligence, refers to intelligence that is applied to analyze, evaluate or critique, and arrive at solutions. Basically, it involves the ability of an individual to identify patterns and accurately predict the outcomes of complex events. This analytical or academic problem-solving intelligence is related to solving problems that require a single solution. It is displayed when the components of intelligence are applied to analyze, evaluate, judge, or compare and contrast. This is what a traditional IQ test measures.

- *Creative Intelligence* is the ability to solve new problems quickly. It is the ability to go beyond what is given to create, invent, and create new solutions. This type of intelligence involves imagination, innovation, and problem-solving. Creativity is demonstrated when someone comes up with a new idea. It may include finding a novel solution to a problem. People are often creative in some domains, but not in others. High creative people usually have intense knowledge about an idea, work on it for some time, look at novel solutions, seek out the advice of others, and take risks. Creativity thinking skills can emerge as both potential and performance factors.
- *Practical Intelligence* (also known as street smarts or common sense) is the ability that individuals use to solve problems faced in daily life. It represents a type of street smarts that is learned from life experiences (especially from mistakes) and cannot be gained from books or formal learning. Learning is perhaps the best way to improve your practical intelligence. Tacit intelligence is an important component of practical (or contextual) intelligence, consisting of knowledge which is required in many real-life situations. Practically gifted individuals are superb in their ability to succeed in any setting. They often become leaders due to their ability to communicate effectively with others. The US Constitution is an example of practical intelligence at work.

The three intelligences or abilities comprise Sternberg's triarchic theory of intelligence. They are illustrated in Fig. 2.2. The three types of intelligence are necessary for success in life and for this reason they may be regarded as "successful intelligence" [6].

The theory of intelligence provides a good way of understanding human intelligence. It carries the idea that intelligences are developing abilities rather than fixed characteristics of an individual. The ability to achieve lifelong success largely depends on a balance of analytical, creative, and practical abilities, a balance which is achieved in order to adapt to, shape, and select environments. Analytical, creative, and practical tests have been used to predict mental and physical health among adults. The triarchic theory emphasizes processes of intelligence, rather than domains of intelligence. Students in the triarchic-intelligence (analytical, creative, practical) condition outperform the other students in terms of the performance assessments.

Intelligence changes with age. IQ is a score obtained by dividing a person's mental age score, obtained by administering an intelligence test, by the person's chronological age, both expressed in terms of years and months. That is, IQ is determined by using the following formula [7]:

$$\text{IQ Score} = \text{mental age}\left(\text{MA}\right) \div \text{chronological age}\left(\text{CA}\right) \times 100$$

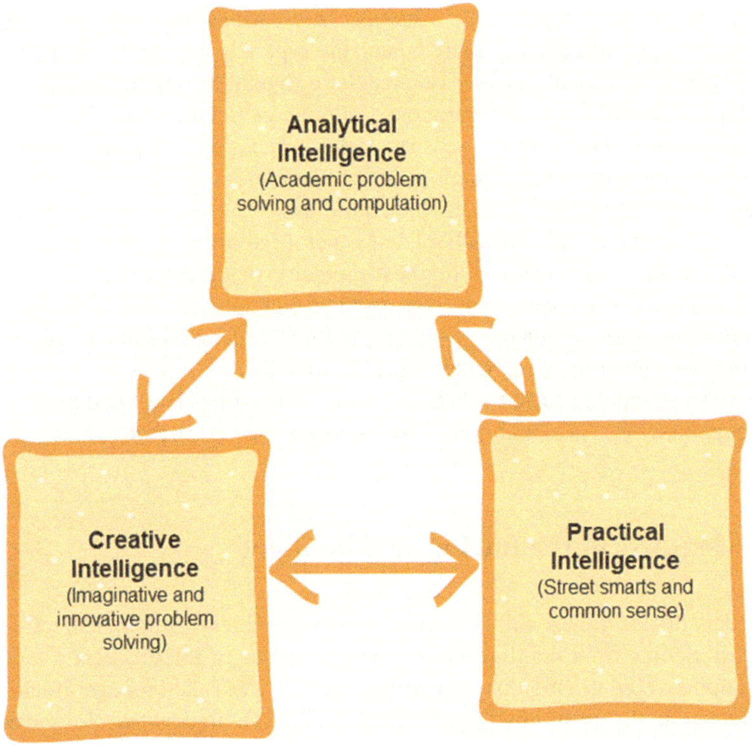


**Fig. 2.2** Sternberg's theory identifies three types of intelligence

Thus an 8-year-old child who does as well as the average 10-year-old child would have an IQ of 125 ($10 \div 8 \times 100$).

## 2.4 Features of General Intelligence

There are some basic assumptions and features of the general intelligence. The key features of general intelligence include [8, 9]:

- General intelligence involves the ability to achieve a variety of goals, and carry out a variety of tasks, in a variety of different contexts and environments.
- A generally intelligent system should be able to handle problems and situations quite different from those anticipated by its creators.
- Real-world systems may display varying degrees of limited generality, but are inevitably going to be a lot more efficient at learning some sorts of things than others.
- Humans display a higher level of general intelligence than existing artificial intelligence programs do, and apparently also a higher level than other animals.

- According to our observations of humans and various theoretical perspectives, the following traits, among many others, are typically associated with generally intelligence: reasoning, creativity, association, generalization, pattern recognition, problem-solving, memorization, planning, achieving goals, learning, optimization, self-preservation, sensory data processing, language processing, classification, induction, deduction, and abduction.
- General intelligence is the ability to acquire basic knowledge and use it in novel situations. People are born with a fixed general intelligence, which can be measured. General intelligence is the best predictor of academic achievement and workplace success across a wide spectrum of domains.
- General intelligence changes with age, is improved by education, and may be hindered by environmental factors such as poverty.
- The more complex a task, the higher it correlates with *g*. The worst performance on cognitive tasks is a better predictor of *g* than the best performance.

## 2.5 Measures of the Intelligence Quotient

Individuals differ in their ability to reason, solve problems, think abstractly, plan, and learn. Scores from intelligence tests are estimates of intelligence. The intelligence quotient (IQ) is a measure of intelligence that is adjusted for age. Intelligence tests are among the most reliable and valid, of all psychological tests for the assessment of intelligence. The IQ is the main determining factor in one's ability to manage formal school education; the ability to learn, process, and repeat information. Over the years, a number of scales of the IQ have been developed.

The IQ test, or the measure of the intelligence quotient, was first developed by the English psychologist Charles Spearman in the first decade of the twentieth century. He measured human intelligence using the general intelligence g factor, which is a psychometric construct that summarizes the correlations observed between an individual's scores on a wide range of cognitive abilities. According to Spearman, general intelligence (or g factor) refers to a general mental ability that underlies multiple specific skills, including verbal, spatial, numerical, and mechanical. These are the skills usually examined to determine the intelligence quotient. These are illustrated in Fig. 2.3.

The g factor typically accounts for 40–50% of the between-individual performance differences on a given cognitive test. IQ scores are often normally distributed, meaning that roughly 95% of the population has IQ scores between 70 and 130. The average score for the test is 100, and any score from 90 to 109 is considered to be in the average intelligence range. Score from 110 to 119 are considered to be High Average. Superior scores range from 120 to 129 and anything over 130 is considered Very Superior or gifted.

Wechsler created the Wechsler Intelligence Scale for Children (WISC) and the Wechsler Adult Intelligence Scale (WAIS) in 1955, with the most up-to-date version being the WAIS-IV. *The most widely used intelligence test for adults* is the Wechsler

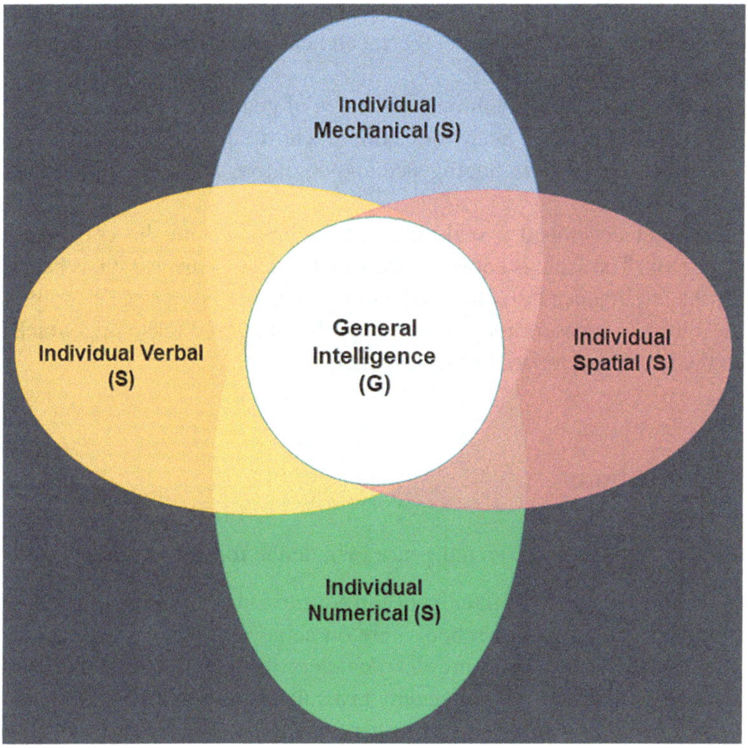


**Fig. 2.3** Spearman's representation of an individual's general intelligence across multiple abilities

Adult Intelligence Scale (WAIS Fourth Edition or WAIS-IV). It comprises tasks that assess several aspects of intelligence: some assess verbal comprehension (e.g., word definition, general knowledge, and verbal reasoning), some assess visuo-spatial reasoning (e.g., puzzle construction, matrix reasoning, and visual perception), some assess working memory (e.g., digit span, mental arithmetic and mental manipulation), and some assess mental-processing speed (e.g., reaction time for detection). This is a comprehensive modern intelligence assessment, standardized on 2200 people ranging from 16 to 90 years of age. It comprises tasks that assess several aspects of intelligence [10]:

- verbal comprehension (e.g., word definition, general knowledge, and verbal reasoning)
- visuo-spatial reasoning (e.g., puzzle construction, matrix reasoning, and visual perception)
- working memory (e.g., digit span, mental arithmetic and mental manipulation)
- mental-processing speed (e.g., reaction time for detection)

The scores on all of these tasks are tallied to calculate a standardized Full Scale Intelligence Quotient (FSIQ) score. IQ researchers sum standardized subtest scores to calculate general intelligence.

One of the widely used traditional measures of general intelligence for children is the Wechsler Intelligence Scale for Children-Third Edition [WISC-III]. This is an IQ test designed to measure intelligence and cognitive ability in children between the ages of 6 and 16.

Alfred Binet developed a scale in 1905 in France with the collaboration of Thomas Simon. The scale is known as the Binet–Simon scale and it has become the basis for the intelligence tests still used today. The Binet–Simon scale comprised 30 items designed to measure judgment, comprehension, and reasoning which Binet deemed the key characteristics of intelligence.

## 2.6 Applications

Some applications of general intelligence include the following [11–13].

- *Education:* The pursuit of intelligence in education is very important. Admission into colleges and graduate schools rely on achievement tests such as the SAT, ACT, GRE, LSAT, etc. These are all barometers of students intelligence. Teaching analytically, creatively, and practically in a cultural setting can make a difference in school achievement. Educators in general have made use of Binet's conception of general intelligence. A student with superior general intelligence will show greater initiative and will possess superior ability to adapt himself.
- *Academic Achievement*: According to research by Robert L. Thorndike, 80–90% of the predictable variance in scholastic performance is due to g. g is closely related to the ability to learn novel material and understand concepts. In elementary school, the correlation between IQ and grades and achievement scores is between 0.60 and 0.70. Achievement test scores are more highly correlated with IQ than school grades. The SAT is primarily a measure of g. In elementary school, the correlation between IQ and grades and achievement scores is between 0.60 and 0.70. In high school, college, and graduate school, the validity coefficients are 0.50–0.60, 0.40–0.50, and 0.30–0.40, respectively. Achievement test scores are more highly correlated with IQ than school grades. This may be because grades are more influenced by the teacher's idiosyncratic perceptions of the student. In a longitudinal English study, *g* scores measured at age 11 correlated with all the 25 subject tests.
- *Job Attainment:* The dispersion of general intelligence scores is smaller in more prestigious occupations than in lower level occupations, suggesting that higher level occupations have minimum g requirements. At the level of individual employees, the association between job prestige and *g* is lower—one large US study reported a correlation of 0.65 (0.72 corrected for attenuation). *g* has a number of correlates in the brain.

- *Job Performance:* Research indicates that tests of *g* are the best single predictors of job performance. g has predictive validity even for the simplest jobs. The hiring process has relied on standard psychometric measures of general intelligence, as predictors of job performance. Research also shows that specific aptitude tests tailored for each job provide little or no increase in predictive validity over tests of general intelligence. It is believed that g affects job performance mainly by facilitating the acquisition of job-related knowledge. General cognitive ability (GCA) remains as the best predictor of job performance. An employee with low general cognitive ability (GCA) will compensate his/her task performance. People high in GCA are capable to learn faster and acquire more job knowledge easily, which allow them to perform better. GCA affects acquisition of job knowledge, which in turn improves job performance. Males generally outperform females in spatial tasks, while females generally outperform males in verbal tasks. This explains the presence of greater numbers of males than females in engineering and technology.
- *Intelligence Tests:* Intelligence tests were originally designed with the practical goal of identifying students in need of alternative education. Many new intelligence tests have developed lately. Undergraduate admission into universities relies on specific aptitude and achievement tests such as the Scholastic Assessment Test (SAT) or the American College Test (ACT). For example, the ACT is strongly g-loaded. Graduate admission requires the Graduate Record Examination (GRE), Graduate Management Admission Test (GMAT), Medical College Admissions Test (MCAT), or the Law School Admission Test (LSAT). These tests have become an integral part of our lives. For example, the Educational Testing Service does not claim that it measures general intelligence, but that the SAT measures verbal and mathematical reasoning abilities.
- *Crime:* Research has shown a correlation between IQ and crime. Children with poor scholastic performance may feel alienated and may be more likely to engage in delinquent behavior, compared to other children who do well. Higher average IQs were very weakly associated with lower levels of property crime, burglary, motor vehicle theft, violent crime, robbery, and aggravated assault. It has also been shown that the effect of IQ is heavily dependent on socioeconomic status and that it cannot be easily controlled away [14].
- *Natural language:* Natural languages are the primary visible forum for intelligence; they are incremental, self-evolving, and self-mutating. Language assimilation and utilization in communication are central to human intelligence. Natural languages and we humans assimilate these. One learns China by growing up in China [15].
- *Music:* People differ in the rate at which they acquire complex skills such as music. Some people develop musical skill far more rapidly than others. There have been several reports of significant associations between general intelligence and skill acquisition in music [16].
- *Creativity*: The creativity of an individual depends on numerous factors, such as knowledge, general intelligence, and emotional intelligence. Does creativity emerge generally or is it domain-specific? Could individuals like Thomas Edison,

Beethoven, and Picasso who became prominent in fields, present creative products in other fields? Creativity thinking skills may emerge as performance factors. It may be regarded as a domain for general everyday skills and it develops into domain-specific forms upon the acquisition of domain relevant skills and knowledge [17].

Other applications of GI include income, occupation, performing artists, intelligent decision-making, online knowledge sharing, brain structure, behavior, and games.

## 2.7 Global General Intelligence

Globalization is a reality in the twenty-first-century workplace. As a result, leaders must function effectively in cross-border situations as well as in domestic contexts. General intelligence is the ability to acquire basic knowledge and use it in novel situations. This section considers how general intelligence is being applied in various nations.

- *United States*: The concept of general intelligence is empirically grounded in the observation of positive correlations between all test scores, as reflected by a general factor termed g. A prevailing notion is that general intelligence emerges from the interaction of multiple processes in the brain. Black–White ability profiles were developed for national standardization sample from the Wechsler Adult Intelligence Scale (WAIS). There was a disparity of approximately one standard deviation between the mean performance of whites over blacks on standardized tests [18].
- *United Kingdom*: Intelligence is one individual characteristic that may be associated with such emotional resilience to inflammatory responses to illness, injury, or stress. Intelligence has been shown to enhance individuals' care of their own health and well-being. There is no sex difference in general intelligence in childhood, but a sex difference favoring males emerges after age 16. There is noticeable sex difference in general intelligence in adults, yet males tend to be overrepresented at both the top and bottom ends of its overall distribution [19].
- *Libya:* Studies in Libya have concluded that there is no gender difference on the Standard Progressive Matrices, a test of nonverbal reasoning that has been used in a large number of countries. Males obtained higher mean scores than females but had lower variance [20]. The tests we used cut to the heart of what intelligence is—ability to adapt to the environment.
- *Jordan:* Social scientists studying authoritarian rule have long recognized the importance of coercive institutions such as armies, national guards, police, and intelligence agencies. Popular policy accounts of Jordan's General Intelligence Directorate (GID) commonly detach it from the country's history, society, and political economy [21]. It has been reported that GID is one of the most important and professional intelligence agencies in the Middle East. Officers serving

in GID are generally appointed by royal decree upon the recommendation of the Director-General; they all have university degrees in different majors and must go through a thorough security check [22].

## 2.8 Benefits

General intelligence is a robust and best predictors of conventional measures of educational achievement, job performance, job satisfaction, training performance, socioeconomic success, social mobility, health, and longevity. Higher general intelligence is often associated with educational and career successes. General intelligence is an important predictor of real-world decision-making. It is strongly related to working-memory capacity or memory span. Intelligent people can solve problems better than less intelligent people. Success in life largely depends on general intelligence. IQ tests are designed to provide a measure of this general ability.

Despite the criticisms and limitations of individually administered intelligence tests, some school districts exclusively rely on standardized IQ test scores to identify giftedness. Traditional measures of general intelligence (e.g., the Wechsler Intelligence Scale for Children) have not been shown to detect many of the problems exceptional children experience. Detroit Tests of Learning Aptitude (DTLA-3) developed by Hammill in 1991 was designed to measure both general intelligence and discrete abilities for children aged 6 through 17 years.

Decades of research have shown that people are born with hereditary potentials for intelligence. Some tend to believe that social, psychological, and economic differences among families create lasting and marked differences in IQ [23]. General intelligence has been shown to enhance individuals' care of their own health and well-being through effective learning skills. It is at the root of our unprecedented technological achievements and is the basis for modern civilization.

## 2.9 Challenges

The concept of general intelligence has not gone without challenges. It seems to be one of those subjects that generate the highest controversies in the modern world. There is considerable debate about what exactly constitutes general intelligence and how best to measure it. Up till today, there is no general consensus on how exactly to define intelligence.

Some critics of general intelligence maintain that the notion of general intelligence is illusory and no single measurement can do justice to the complexity of people's mental abilities. They strongly oppose the idea that intelligence was heritable, innate, and could be relegated to a single number. They allude to the fact that no such global mental capacity exists and that apparent "intelligence" is really just a by-product of one's opportunities to learn skills. For example, the Educational

Testing Service does not claim that it measures general intelligence, but that the SAT measures verbal and mathematical reasoning abilities [24].

The standard IQ test was not a great predictor of one's success because it can only measure a narrow range of analytical abilities. It has been argued that IQ tests are biased in favor of white, middle-class people. Since the majority of the world's population does not live in Europe or North America, the cultural specificity of the IQ tests is crucial. Some psychologists argue that intelligence is genetic, or inherited, while others claim that it is largely a general ability which can be developed.

## 2.10 Conclusion

General intelligence is the measure of cognitive ability that has received more attention than any other form of intelligence. It relates to the knowledge and faculties an individual needs for skill acquisition. It is one of the most pervasive concepts in psychology. The existence of GI as a single measure of intellectual ability is orthodoxy among psychologists and the general public. General intelligence is a general mental capability to reason, think abstractly, solve problems, plan, and learn across domains. The nature of general intelligence has been studied by many investigators. Individuals with high general intelligence (or high g scores) will be those usually performing well. Intelligence changes with age and is improved by education. For more information about general intelligence, one should consult the books in [25–33] and related journals:

- *Intelligence*
- *Science*
- *The American Journal of Psychology*
- *British Journal of Psychology*
- *Psychological Inquiry*

## References

1. S.M. Reader, Y. Hager, K.N. Laland, The evolution of primate general and cultural intelligence. Philos. Trans.Royal Soc. B Biol. Sci. **366**(1567), 1017–1027 (2011)
2. M.N.O. Sadiku, S.M. Musa, A. Ajayi-Majebi, The essence of general intelligence. Int. J. Trend Res. Develop. **7**(6) (2020)
3. P. Shrestha, General intelligence. Psychestudy **18** (2017) https://www.psychestudy.com/cognitive/intelligence/general
4. C. Ruhl, Intelligence: definition, theories and testing (2020), https://www.simplypsychology.org/intelligence.html#:~:text=Intelligence%20has%20been%20defined%20in,demands%20of%20the%20environment%20effectively
5. T. Donvito, What is practical intelligence and can it be improved? (2019), https://www.rd.com/article/practical-intelligence/

6. Intelligence and creativity, https://courses.lumenlearning.com/wmopen-psychology/chapter/what-are-intelligence-and-creativity/
7. 10.1 Defining and measuring intelligence, https://opentextbc.ca/introductiontopsychology/chapter/9-1-defining-and-measuring-intelligence/
8. B. Goertzel, Artificial general intelligence (2015), http://www.scholarpedia.org/article/Artificial_General_Intelligence
9. N. Paz-Baruch, Educational and learning capital as predictors of general intelligence and scholastic achievements. High Abil. Stud. **31**(1), 75–91 (2020)
10. J. Dubois et al., A distributed brain network predicts general intelligence from resting-state human neuroimaging data. Philos. Trans. Royal Soc. B Biol. Sci. **373**(1756) (2018)
11. g factor (psychometrics), Wikipedia, the free encyclopedia
12. W. Champlin, Dynamic personality traits and general intelligence, Masters Thesis, University of Southern California, 1940
13. P. Carney, General intelligence or g and it's impact on education (2016), https://www.paulcarneyarts.com/single-post/2016/04/20/General-Intelligence-or-g-and-it%E2%80%99s-impact-on-education
14. Intelligence quotient, Wikipedia, the free encyclopedia https://en.wikipedia.org/wiki/Intelligence_quotient
15. S. Iyengar, T. Rabe, A general language framework for general intelligence, in *Proceedings of International Conference on Artificial Intelligence* (2015), pp. 347–353
16. A.P. Burgoyne, L.J. Harris, D.Z. Hambrick, Predicting piano skill acquisition in beginners: the role of general intelligence, music aptitude, and mindset. Intelligence **76** (2019)
17. F. Şahin, General intelligence, emotional intelligence and academic knowledge as predictors of creativity domains: a study of gifted students. Cogent Education **3**(1), 1218315 (2016)
18. S.M. Kaiser, Ability patterns of black and white adults on the Wechsler adult, Doctoral Dissertations, Texas A&M University, 1986
19. W. Johnson, A. Carothers, I.J. Deary, Sex differences in variability in general intelligence: a new look at the old question. Perspect. Psychol. Sci. **3**(6), 518–531 (2008)
20. A. Al-Shahomee, Gender differences on general intelligence in Libya. Mank. Q. **60**(1), 93–98 (2019)
21. P.W. Moore, A political-economic history of Jordan's General Intelligence Directorate: authoritarian state-building and fiscal crisis. The Middle East Journal **73**(2), 242–262 (2019)
22. General Intelligence Directorate (Jordan), Wikipedia, the free encyclopedia https://en.wikipedia.org/wiki/General_Intelligence_Directorate_(Jordan)
23. L.S. Gottfredson, The general intelligence factor, https://www1.udel.edu/educ/gottfredson/reprints/1998generalintelligencefactor.pdf
24. S.D. Ridgell, J.W. Lounsbury, Predicting academic success: general intelligence, 'Big Five' personality traits, and work drive. Coll. Stud. J. **38**(4) (2004)
25. Na, *General Intelligence and Test of Reasoning* (S. Chand Publishing, 2017)
26. B. Goertzel, C. Pennachin, N. Geisweiller, *Engineering General Intelligence, Part 1: A Path to Advanced AGI via Embodied Learning and Cognitive Synergy* (Atlantis Press, 2014)
27. N.J. Mackintosh, *IQ and Human Intelligence* (Oxford University Press, 1998)
28. A. Binet, T. Simon, C.H. Town, *A Method of Measuring the Development of the Intelligence of Young Children*, 3rd edn. (Chicago Medical Book, Chicago, IL, 1915)
29. M.A. Woodley, A.J. Figueredo, *Historical Variability in Heritable General Intelligence: Its Evolutionary Origins and Socio-Cultural Consequences* (University of Buckingham Press, 2013)
30. V. Experts, *General Intelligence and Test of Reasoning*, 2nd edn. (S. Chand, New Delhi, 2017)
31. H. Nyborg (ed.), *The Scientific Study of General Intelligence: Tribute to Arthur Jensen* (Pergamon, 2003)
32. K. I. S. Rathore (ed.), *General Intelligence & Reasoning* (Rathore Publishers & Distributors, 2015)
33. R. J. Sternberg, E. L. Grigorenko (eds.), *The General Factor of Intelligence; How General Is it?* (Lawrence Erlbaum, 2002)

# Chapter 3
# Naturalistic Intelligence

**Contents**



> Naturalism teaches one of the most important things in this world. There is only this life, so live wonderfully and meaningfully.
>
> —Greg Graffin

## 3.1 Introduction

The idea of intelligence was first introduced by Francis Galton in 1885. Intelligence or cognitive development is a bio-psychological potential to process information that can be activated to solve problems. It is the ability to learn from experience and to adapt to, shape, and select environments. It has traditionally been thought of as an inherited entity that one is born with and cannot be changed.

Professor Gardner at Harvard's Graduate School of Education, the founding father of the multiple intelligences theory, argued that there are better or alternative ways to measure intelligence than standard IQ tests. He first proposed nine intelligences and insists that all people are born with one or more intelligences and that people can be smart in a number of different ways. This implies that we are all intelligent in different ways. The best way to develop these intelligences is by practicing skills in their respective areas. Naturalistic intelligence (or naturalist intelligence) is one of Howard Gardner's nine multiple intelligences, which involves how sensitive a person is to nature and the world [1].



M. N. O. Sadiku, S. M. Musa, *A Primer on Multiple Intelligences*,
https://doi.org/10.1007/978-3-030-77584-1_3

This chapter provides an introduction on naturalistic intelligence. It begins by defining naturalistic intelligence (NI). It presents some of the characteristics of NI and how it can be measured. It provides some tips for developing naturalistic intelligence and some quotes related to naturalistic intelligence. It highlights some benefits and challenges of NI. The last section concludes with comments.

## 3.2 What Is Naturalistic Intelligence?

In 1983, Howard Gardner published seven types of intelligence in his famous book [2]. These are: verbal/linguistic, mathematical/logical, spatial, musical, kinesthetic, interpersonal, and intrapersonal. Figure 3.1 illustrates Howard Gardner's theory of multiple intelligences. In 2006 (23 years later), Gardner added an eighth branch to his model in his another book [3]. This eighth branch is known as naturalistic intelligence. Naturalistic intelligence is a new addition that meets Gardner's criteria as an intelligence. Gardner accurately pinpointed parts of the brain that are correlated to each of his intelligences. Naturalistic intelligence is seated in the parts of the brain responsible for sensing patterns and for acute sensory perceptions.

Naturalistic intelligence is one of the nine multiple intelligences proposed by researcher Howard Gardner. Naturalistic intelligence refers to a person's interest in the "natural" world of animals, plants, and the natural work around them. It is the

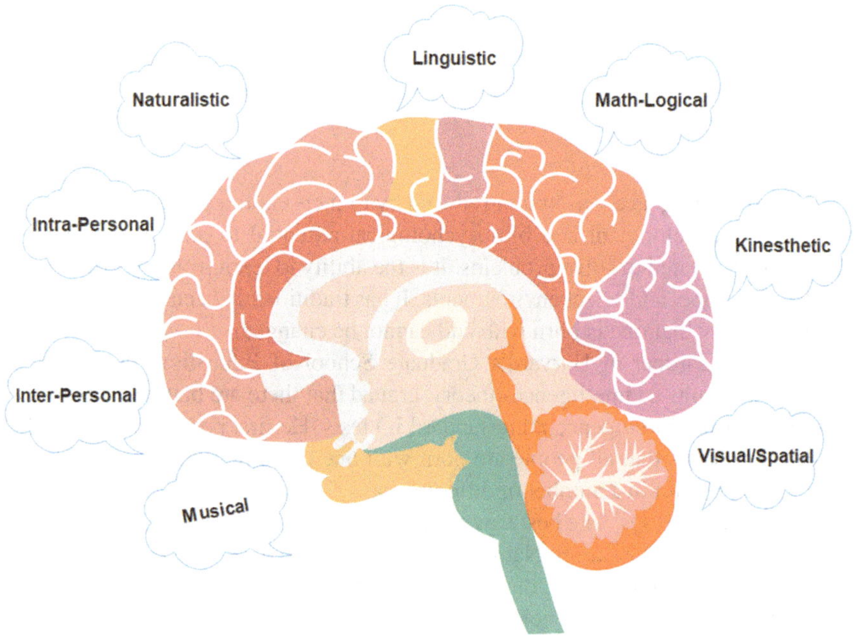


**Fig. 3.1** Howard Gardner's theory of multiple intelligences

ability to identify and classify patterns in natural environment. Individuals with naturalist intelligence have a sensitivity to and appreciation for nature. Naturalists are inspired and rejuvenated by nature.

According to Gardner, naturalistic intelligence is the ability to identify, classify, and manipulate elements of the environment, objects, animals, or plants. The intelligence has to do with being in tune with nature and one's natural environment. It can be seen in weather forecasting, gardening, cooking, civil engineering, landscape design, and plant/animal classifications. Like other intelligences or strengths, naturalistic intelligence is not fixed like a standard IQ. It can grow and develop throughout a person's life. Figure 3.2 illustrates naturalistic intelligence.

A naturalistic person or naturalist is someone who possesses high naturalistic intelligence. Naturalists learn best when the subject is related to something prominent in nature. They have the sense to appreciate life and its wonders. They are "nature smart." They show interest in natural elements, like gardens, forests, mountains, etc. They have an affinity for animals, plants, vegetables, and fruit. They have an uncanny ability to make observations and distinctions about nature. They may petition their local politicians, or work with others to create green spaces in their communities.

People with high naturalistic intelligence often find their way into careers that focus on environmental science, botany, biology, ecology, astronomy, zoology, agriculture, horticulture, archaeology, volcanology, oceanography, geology, paleontology, meteorology, or forensic science. Historical figures with the highest naturalistic intelligence include [4]:

- *Charles Robert Darwin:* This is history's most famous evolutionary scientist. He was a British geologist, known for contributing to the theory of evolution. He proposed that all life evolved from the common ancestors.

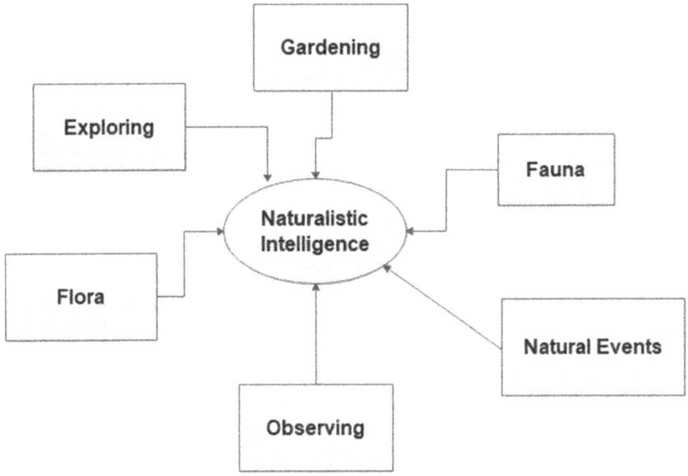


**Fig. 3.2** An illustration of naturalistic intelligence

- *George Washington Carver*: The famous American botanist, known for inventing peanut butter and innumerable things from peanuts.
- *Rachel Carson:* An American conservationist and marine biologist. She wrote a book about the damages that pesticides were causing to the environment.
- *Luther Burbank:* Renowned American agricultural scientist and horticulturist, who developed more than 800 plant varieties.
- *Alexander von Humboldt:* A famous naturalist and explorer, the first person to suggest that humans were having an impact on the natural world and causing climate change.
- *E.O. Wilson:* The world's greatest naturalist and the father of sociobiology.
- *William Wordsworth*: He integrated naturalistic intelligence into his poetry.

Others include Jacques Cousteau, Carl Sagan, Al Gore, David Suzuki, Jane Goodall, Steve Erwin, and Neil deGrasse Tyson. Figure 3.3 shows a typical naturalist [4].

Naturalistic intelligence should not be confused with natural intelligence, which offers insight about how to equip machines with human-like capabilities. Natural intelligence uses AI in building devices that get the job done. Also, naturalistic intelligence is not worshipping nature, as some suggest. There is a significant relationship between naturalist intelligence and environmental attitude. Some regard naturalistic intelligence as environmental awareness.

**Fig. 3.3** A typical naturalist [4]

## 3.3 Characteristics of Naturalist Intelligence

Individuals with high naturalistic intelligence show an inclination towards the natural world. They have the following unique characteristics [5, 6]:

- They express a desire to understand how things work.
- They care about the environment and like to be in touch with nature.
- They like to explore and discover new species and behaviors.
- They are interested in using tools to help observation—microscopes, binoculars, telescopes.
- They love subjects related to nature, such as botany, zoology, archaeology, and geology.
- They love learning about the geographical changes—the tides, the climate, volcanoes, earthquakes, tornadoes, etc.
- They specialize as gardeners, farmers, scientists, ecologists, or veterinarians.
- They are sensitive to and appreciative of nature.
- They enjoy gardening and feel alive when in contact with nature.
- They have a hobby that involves nature.
- They are environmentally conscious.
- They hate anything that destroys the planet such as pollution and cutting of trees.
- They participate in environmental or nature preservation causes.
- They care for animals and/or plants and enjoy having pets.
- They read nature books or magazines.
- They may be interested in joining the Boy or Girl Scouts.

Besides these general characteristics, the characteristics related to naturalistic intelligence can also be divided in left and right brain strengths, left being the more analytic. Characteristics related to the right side of the brain are [7]:

- Feels at their best in the outdoors
- Strives for balance with nature and mind and body
- Demonstrates an empathy with nature and its creatures
- Has a strong sense of responsibility towards the environment
- Possesses a sensitivity to animal abuse and environmental destruction
- Enjoys exploration, adventure, open-ended experiences
- Feels an affinity towards animals in general, pets in particular

Characteristics related to the left side of the brain are [7]:

- Analyses of information
- Has an eye for detail
- Sees patterns
- Identifies categories
- Reasons in hierarchy (can rank items by significance and relationship)
- Memorizes schematically
- Looks for structure
- Understands statistics

- Prefers charts, timelines
- Has a strong interest in ecology, biology, or chemistry

Through these characteristics, one can identify individuals with naturalistic intelligence.

## 3.4 Measurement of Naturalistic Intelligence

Armstrong has used the following ten statements to measure naturalistic intelligence in his "Multiple Intelligences Profile Assessment" [8]:

1. I write well and enjoy putting thoughts on paper (or in the computer).
2. I enjoy telling stories or jokes.
3. I enjoy word games like Scrabble or Anagrams.
4. I enjoy reading books and magazines.
5. I have a good vocabulary.
6. I enjoy communicating by talking or writing.
7. Malay/English and histories were among my favorite subjects in school.
8. When I walk down the pathway in town, I pay more attention to the words on billboards than to the people around.
9. I enjoy entertaining others or myself with tongue twisters, nonsense rhymes, limericks, or puns.
10. I get more out of listening to the radio or a spoken-word cassette than I do from television or films.

## 3.5 Developing Naturalist Intelligence

As children grow, it is necessary to ensure that the naturalistic intelligence contributes to their attitudes and behaviors towards the environment. Children who are "nature smarts" may enjoy TV shows, games, and stories that deal with nature, animals, or natural phenomena. They enjoy being outside playing, looking, seeking, gardening, camping, hiking, climbing, exploring, and getting their hands dirty. Children with high naturalistic intelligence tend to ask endless questions about what they observe.

Naturalist intelligence can be developed in childhood or later in life. It can be developed by using appropriate teaching methods. Activities that a teacher might use to reach students should focus on direct observation and exploration of the natural world. Help them think through how what they are learning impacts the world. Since naturalistic intelligence is much a hands-on experience, the best way to develop naturalistic intelligence is by getting outside. Go outside, appreciate nature, and admire the trees, the grass, the flowers, etc. You may take a field trip. For example, if a person moves from an urban area to a rural area, he might become exposed to fields, hiking, and mountains and develop his natural intelligence. The following tips may also help in developing naturalistic intelligence [9]:

- Go on a nature walk
- Cultivate a garden
- Be involved in hands-on activities
- Read books and articles about nature and the environment
- Visit various natural locations and places where they can mingle with nature (including farms)
- Keep a Nature Journal and create observation notebooks
- Describe changes in the local or global environment that affect you
- Become involved in the care for pets, wildlife, gardens, or parks
- Use scientific gadgets such as binoculars, telescopes, microscopes, or magnifiers to see things differently
- Draw or photograph natural objects
- Collect natural organisms—feathers, leaves, flowers, etc.
- Celebrate Earth Day on April 22nd.
- Write articles about nature (poems, stories, articles, blogs, etc.)

You can use these tips to help your child or student excel in school.

As a result of participating in the activities suggested above, the following skills will be developed [10, 11]:

1. *Observation Skills:* This is probably the most fundamental skill of naturalist intelligence. Interacting with the nature helps develop observation skills.
2. *Empathy:* Getting involved in activities that lead to interaction with the environment will help in developing empathy towards people, animals, and nature.
3. *Sensitivity:* Environmental sensitivity helps to develop concern for the environment and helps the students understand concepts like sustainability and efficiency.
4. *Curiosity:* Naturalists are known to always ask questions and are curious or inquisitive about the environment. Curiosity is the first step in making new discoveries outside.
5. *Motor Skills:* Engaging the naturalists in activities that involve collecting, sorting, and discovering will help in developing motor skills and will keep the mind active.

## 3.6 Applications

The naturalist intelligence deals with how we relate to our surroundings and where we fit into it. It is being in tune with nature and our natural surroundings. Naturalist intelligence is paramount in the era of technology and globalization because the world needs people who care for and maintain the environment that is safe, clean, and comfortable to stay healthy. Although there are many ways we can grow or apply naturalistic intelligence, here we consider educational and environmental implications of naturalist intelligence.

- *Education:* In education, efforts are required to develop intelligence in students. Naturalist intelligence can be developed in childhood or later in life, just like other intelligences. Schools should be teaching all the nine Gardner's multiple intelligences. There are the students who exhibit high naturalist intelligence, while others are low in naturalistic intelligence. Teachers can help all students develop NI and become "nature smarts." They can broaden the curriculum to include teaching all the intelligences. Figure 3.4 shows how children learn about naturalistic intelligence [7]. Strategies for teaching naturalist intelligence include relating, discovering, observing, planting, collecting, sorting, comparing, and displaying. Students with naturalist intelligence are those who seem to be in love with the natural world. Students demonstrate naturalist intelligence when they [12]:
  - are very comfortable outdoors
  - are aware of their natural surroundings
  - feel a definite sense of connection to the rest of nature

**Fig. 3.4** Children learn about naturalistic intelligence [7]

- recognize patterns and colors
- are good at sorting and classifying
- have keen observational skills and observe patiently
- understand and can explain natural phenomena
- nurture living things, through gardening or taking care of pets or bringing home stray animals
- choose to read books and watch programs about animals and ecosystems
- seek music related to nature
- readily follow cyclic patterns in nature such as tides, seasons, moon phases, and climate
- enjoy collections of rocks, leaves, flowers, shells, feathers, etc.
- name zoos, farms, wildlife parks, aquariums, and pet stores as "special places" for them
- enjoy recreational activities in the outdoors, such as hiking, rock climbing, cross-country skiing, camping, sailing, scuba diving, etc.

When students are exposed to these naturalist intelligence activities, they become nature smart. Multiple intelligence learning techniques must also be included to help students become well-rounded.

- *Environmental Awareness:* This is a tendency to behave in a pro-environment manner. There is a relationship between naturalistic intelligence with environmental awareness. Environmental awareness is the emotional level and commitment to environmental issues, while naturalistic intelligence is the ability to understand the natural environment. Naturalist intelligence may be regarded as the key to understanding the living environment. It is well known that human activities have been harmful and cause changes to the environment. Humans have not been good stewards or managers of the environment. It is expedient to ensure that the naturalistic intelligence contributes to the positive attitudes and behaviors of students towards the environment. The students who are high in naturalistic intelligence understand environmental problems, enjoy outdoor activities, have scientific hobbies, and are concerned for the environment [13]. Children should be taught to act pro-environmentally, whether locally or globally.

Other applications of naturalistic intelligence include learning grammar and human ecology.

## 3.7 Quotes About Natural Intelligence

The following quotes related to naturalistic intelligence will help your critical thinking [14–16]:

> Intelligence without ambition is a bird without wings.—Salvador Dali.

> Nature is not a place to visit. It is home.—Gary Snyder

The garden suggests there might be a place where we can meet nature halfway.—Michael Pollan

The earth has music for those who listen.—George Santayana

Adopt the pace of nature: Her secret is patience.—Ralph Waldo Emerson

The human race is challenged more than ever before to demonstrate our mastery, not over nature but of ourselves.—Rachel Carson

I want my children to understand the world, but not just because the world is fascinating and the human mind is curious. I want them to understand it so that they will be positioned to make it a better place—Howard Gardner

Both science and history are moving targets. Scholars in the twenty-first century are much more aware than those of earlier generations that scientists operate under the influence of powerful metaphors (science as exploration, discovery, documentation, thrust and counter-thrust), and that both the scope and the tools of history undergo continual changes.—Howard Gardner

There is no such thing as free, nothing is free, so please don't steal, especially from the future—Environmental Sayings.

Earth and sky, woods and fields, lakes and rivers, the mountain and the sea, are excellent schoolmasters, and teach some of us more than we can ever learn from books.—John Lubbock

From the naturalistic point of view, all men are equal. There are only two exceptions to this rule of naturalistic equality: geniuses and idiots.—Mikhail Bakunin

Either everything in man can be traced as a development from below, or something must come from above. There is no avoiding that dilemma: you must be either a naturalist supernaturalist.—T. S. Eliot

Naturalism teaches one of the most important things in this world. There is only this life, so live wonderfully and meaningfully.—Greg Graffin

## 3.8 Benefits and Challenges

Naturalistic intelligence is the intelligence that deals with the natural, the environment of residence, or anything related to nature. Speculation has it that naturalistic intelligence helped our ancient hunter-gatherer ancestors to identify which flora and fauna were edible and which ones were not. People who have naturalistic intelligence are the ones who wonder about the things they see around them and ask endless questions about what they observe. They make astute observations about natural changes, emerging patterns, and natural phenomena. They love books, shows, or videos about nature or natural phenomena, or animals. They often find their way into professions that deal with or are focused on biology, agriculture, horticulture, and archaeology [17].

Children displaying "nature smarts" are often interested in human behaviors, or the behaviors, habits, or habitats of other species. They may enjoy subjects, shows, and stories that deal with animals or natural phenomena. They show unusual interest in subjects like biology, zoology, botany, geology, meteorology, paleontology, or astronomy [17].

Some critics have considered naturalistic intelligence to be more of an ability than an intelligence. Why is it that the notion of intelligence is too often associated with its opposite, stupidity? White supremacists (some of whom are doctors, educations, and scientists) have historically labeled minority groups "stupid." Measuring skulls in Nazi Germany "proved" Jews' inferiority. Psychometricians often deny that the subjectivity of the researcher, the prejudices of the researcher, are built into the test. Testing is not innocent. Some find the notion of multiple intelligences superficial, reductionistic, and naïve [18].

## 3.9 Conclusion

Naturalistic intelligence is one of the nine multiple intelligence proposed by Gardner. It is the ability to understand the natural environment well and enjoy nature, and use that ability productively. It is being in tune with nature and one's natural surroundings. It enables human beings to recognize, categorize, and draw upon certain features of the environment. It is related to how we connect and adjust to our natural surroundings. It is the intelligence we all share. People with high naturalistic intelligence usually feel most at home when they are outdoors in nature. This is the best way for naturalistic learners to absorb information. More information about naturalistic intelligence can be found in the books in [2, 3, 12, 19–21] and in the journals related to it: *Intelligence, Journal of Intelligence,* and *Journal of Educational Psychology.*

## References

1. M.N.O. Sadiku, T.J. Ashaolu, S.M. Musa, Naturalistic intelligence. Int. J. Scienti. Adv. **1**(1), 1–4 (2020)
2. H.E. Gardner, *Frames of Mind: The Theory of Multiple Intelligence* (Basic Book, New York, 1983)
3. H.E. Gardner, *Multiple Intelligences: New Horizons in Theory and Practice* (Basic Books, 2006)
4. Vital characteristics and examples of naturalistic intelligence, https://psychologenie.com/characteristics-examples-of-naturalistic-intelligence
5. Naturalistic intelligence, the ability to understand the environment, https://www.iberdrola.com/talent/naturalistic-intelligence

6. Naturalist intelligence, https://www.russellrowe.com/multiple-intelligences/naturalist.pdf
7. Naturalistic intelligence, International Montessori Schools and Child Development Centres Brussels, Belgium, https://www.international-montessori.org/wp-content/uploads/2014/03/Naturalistic-Intelligence.pdf
8. N.S. Mumthas, T.K.U. Farooque, Measuring naturalistic intelligence – A contextual and visual approach (2012), https://www.academia.edu/9795294/Measuring_Naturalistic_Intelligence_A_Contextual_and_Visual_Approach_Paper_Submitted_for_National_Seminar_On_Naturalistic_intelligence
9. Naturalist intelligence learning from Inspiring Breakthrough, http://www.inspiring-breakthrough.co.uk/learning-styles/naturalist-learning.htm
10. Naturalist intelligence, https://greycaps.com/theteacher/Community/Naturalist-Intelligence
11. 9 Essential skills of naturalist intelligence (and how they help you), https://nature-mentor.com/naturalist-intelligence-skills/
12. J. Glock, S. Wertz, M. Meyer, *Discovering the Naturalist Intelligence: Science in the Schoolyard* (Chicago Review Press, 1998)
13. Z.B. Ningrum, T.E.B. Soesilo, H. Herdiansyah, Naturalistic intelligence and environmental awareness among graduate students, in *E3S Web of Conferences* (2018)
14. T.O. Kowald, Educational outdoor activities to build nature smarts (2014), https://www.connectionsacademy.com/support/resources/article/educational-outdoor-activities-to-build-nature-smarts
15. Quotes about natural intelligence, https://www.quotemaster.org/natural+intelligence
16. Howard Gardner quotes, https://www.goodreads.com/author/quotes/37381.Howard_Gardner
17. L.O. Wilson, The eighth intelligence – Naturalistic intelligence, https://thesecondprinciple.com/optimal-learning/naturalistic-intelligence/#:~:text=General%20Description%3A%20Naturalist%20intelligence%20deals,or%20habitats%20of%20other%20species
18. M. Morris, Chapter 8: the eighth one: naturalistic intelligence. Counterpoints **278**, 159–173 (2004)
19. H.E. Gardner, *Intelligence Reframed: Multiple Intelligences for the 21st Century* (Basic Books, New York, 1999)
20. K.P. Roth, *Naturalist Intelligence: An Introduction to Gardner's Eighth Intelligence* (Corwin, 1999)
21. W.I. Loh, G. Jacobs, *Nurturing the Naturalist Intelligence* (Kagan Cooperative Learning, 2003)

# Chapter 4
# Social Intelligence

**Contents**



Our intelligence is what makes us human, and AI is an extension of that quality.

—Yann LeCun

## 4.1 Introduction

Intelligence refers to practical problem-solving ability, verbal ability, social competence, and effective adaptation to one's environment [1]. This ability exclusively distinguishes humans from the rest of beings in the animal kingdom. It emerges from local interactions among a large number of individuals, none of whom acts as a leader or central controller.

In the past, scores from IQ tests were regarded as the standard in defining how successful an individual would be in life. Individuals who score high on IQ tests are regarded as having intellectual giftedness. Intelligence measured by IQ tests is not the be all and end all. Two kinds of intelligences that are unmeasurable in IQ tests are pivotal for success in life and career: Emotional Intelligence (EI) and Social Intelligence (SI). Social intelligence is different from academic intelligence. For example, some individuals who are successful in solving problems may find it difficult getting along with others [2].


M. N. O. Sadiku, S. M. Musa, *A Primer on Multiple Intelligences*,
https://doi.org/10.1007/978-3-030-77584-1_4

This chapter provides an introduction on social intelligence (SI) and its ramifications. It begins by explaining what SI is all about and its relationship with other types of intelligence. It addresses the key elements of social intelligence. It presents some demonstration and applications of SI. It discusses how to measure social intelligence. It highlights some benefits and challenges of social intelligence. The final section concludes with comments.

## 4.2 What Is Social Intelligence?

The concept of social intelligence, also known as interpersonal intelligence or social sensitivity, was introduced in 1920 by American psychologist Edward Thorndike, who divided intelligence into three facets pertaining to the ability to understand and manage ideas (abstract intelligence), concrete objects (mechanical intelligence), and people (social intelligence). Thorndike characterized social intelligence as the ability to accomplish interpersonal tasks.

Social intelligence is the ability to understand and manage men and women, boys and girls, and to act wisely in human relations or social situations. It is the ability to read nonverbal cues or make accurate social inferences. It is about figuring out the best way to get along with others. It is the ability to adequately understand and evaluate one's own behavior and the behavior of others. It is the ability to get along well with others and win their cooperation. It is the key to life and career success. Social intelligence is equivalent to interpersonal intelligence, one of nine multiple intelligences proposed by Howard Gardner.

Social Intelligence (SI), or street smarts or people skills, is about understanding your environment and having a positive influence in your social interactions. It is an adaptation for dealing with highly complex social situations, such as politics, romance, family relationships, quarrels, and collaboration [3]. No one is born socially intelligent. SI is mostly learned from experience with people and from success and failures in social settings. We are wired to connect and our manners are essentially cultural social intelligence.

Unlike general intelligence, social intelligence can be developed. Social intelligence increases with academic level and experience of a person. Cognitive intelligence, emotional intelligence, and social intelligence are important components of general/human intelligence. Although social intelligence and emotional intelligence are different from cognitive intelligence, social intelligence is strongly related to emotional intelligence, political skill, and empathy. Empathy is the awareness of another person's thoughts, feelings, and intentions [4]. Emotional intelligence is a set of emotional and social abilities, competencies, and skills that enable individuals to cope with daily demand. It is the ability to identify and manage your one's emotions and the emotions of others. It is significantly related to the ability to exercise personal judgment in decision-making. Emotional intelligence and social intelligence provide a valuable understanding why some people behave more intelligently

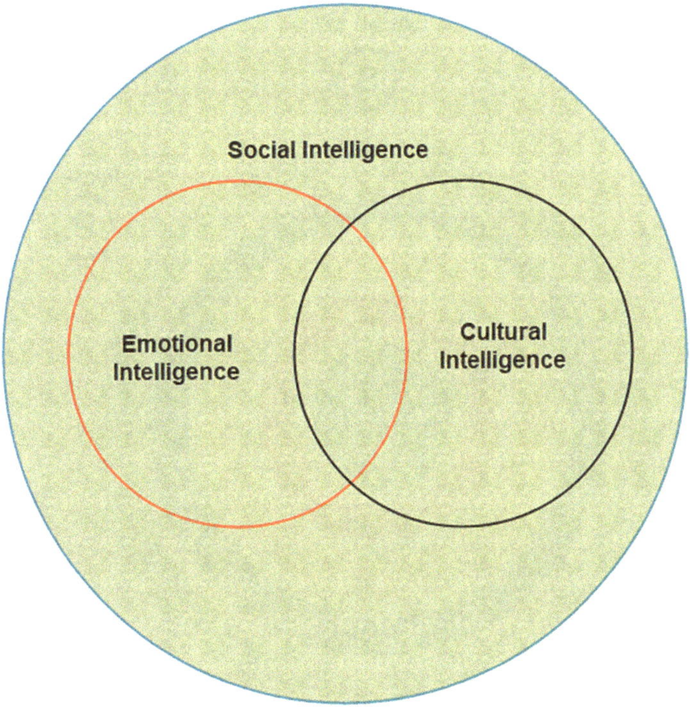


**Fig. 4.1** Relationship between social intelligence, emotional intelligence, and cultural intelligence

than others. Cultural intelligence builds on emotional intelligence and social intelligence. The three are related as shown in Fig. 4.1.

## 4.3 Elements of Social Intelligence

Social intelligence is about successfully navigating interpersonal relationships. Individuals with high social intelligence can sense how others feel and intuitively know what to say in social settings. The social intelligence profile is shown in Fig. 4.2. The key elements of social intelligence include [5]:

1. Verbal fluency and conversational skills.
2. Knowledge of social roles, rules, and scripts.
3. Effective listening skills.
4. Understanding what makes other people tick.
5. Role playing and social self-efficacy.
6. Impression management skills.

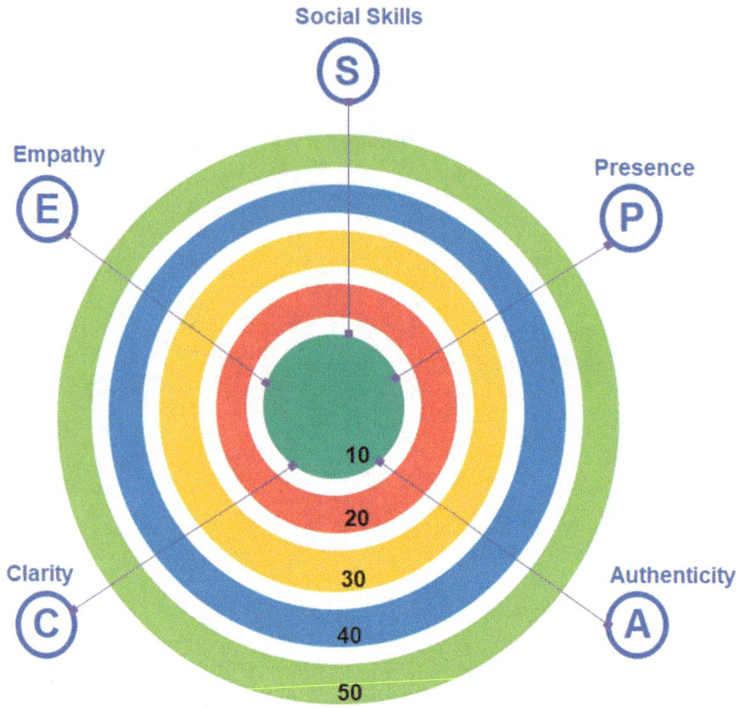


**Fig. 4.2** Social intelligence profile

The core characteristics of someone who is socially intelligent include [6, 7]:

1. They do not try to elicit a strong emotional response from anyone they are holding a conversation with.
2. They do not speak in definitives about people, politics, or ideas.
3. They don't immediately deny criticism, or have such a strong emotional reaction to it that they become unapproachable or unchangeable.
4. They do not confuse their opinion of someone for being a fact about them.
5. They never overgeneralize other people through their behaviors. They rarely use absolutes. They don't use "you always" or "you never" to illustrate a point.
6. They speak with precision and choose their words carefully.
7. They know how to practice healthy disassociation.
8. They do not try to inform people of their ignorance.
9. They validate other people's feelings.
10. They recognize that their "shadow selves" are the traits, behaviors, and patterns that aggravate them about others.
11. They do not argue with people who only want to win, not learn.
12. They listen to hear, not respond.

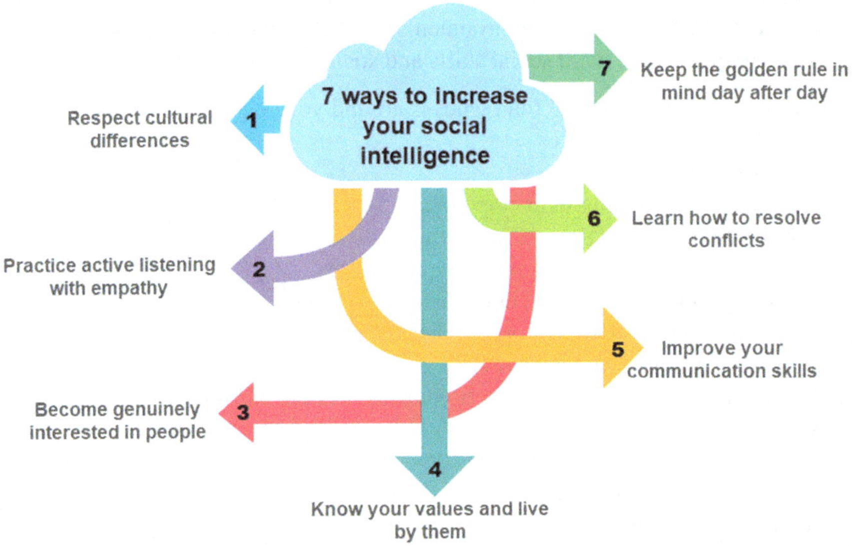


**Fig. 4.3** Five ways to increase your social intelligence

13. They do not post anything online they would be embarrassed to show to a parent, explain to a child, or have an employer find.
14. They do not consider themselves a judge of what is true.
15. They do not "poison the well," or fall for ad hominem fallacy to disprove a point.
16. Their primary relationship is to themselves, and they work on it tirelessly.
17. They recognize mistakes and are ready to accept apologies and forgive.
18. They do not immediately deny criticism.
19. They value compassion over empathy.
20. They see problems as growing opportunities, not finalities.

## 4.4 Improving Social Intelligence

Humans are social beings. Since SI is simply the ability to get along with people and act wisely in social situations, individuals can learn as they grow up, mature, and gain experience in dealing with others. SI develops from experience with people and learning from one's success and failures in social settings. Improving SI takes deliberate effort and hard work in changing your behavior.

One must first develop a personal vision for change and identify areas of social weakness and strength. One must work on becoming a better speaker and developing basic communication skills.

Keep in mind that social intelligence is the skill of observing, interpreting, connecting, and adapting in social environments. With that in mind, the following strategies can help a person build social skills and social intelligence [8, 9]:

- Pay close attention to what (and who) is around you.
- Respect cultural differences.
- Don't suppress empathy.
- Practice active listening with empathy.
- Appreciate the important people in your life.
- Make quick connections with new acquaintances.
- Become genuinely interested in people.
- Seek opportunities to encourage and help others.
- Know your values and live by them.
- When you feel fear or anger, take time to slow down and reflect.
- Assume the best in others.
- Improve your communication skills.
- Learn how to resolve conflicts.
- Invest in your relationships.
- Realize that attention is the currency of all relationships.
- Pay attention to other people.
- Smile more often.
- Keep the golden rule in mind day after day.

Some of the ways to increase your social intelligence are shown in Fig. 4.4. Like any skill, one must put in the necessary time reading, listening, and practicing in developing social intelligence. Improving your social intelligence will make you become socially smarter.

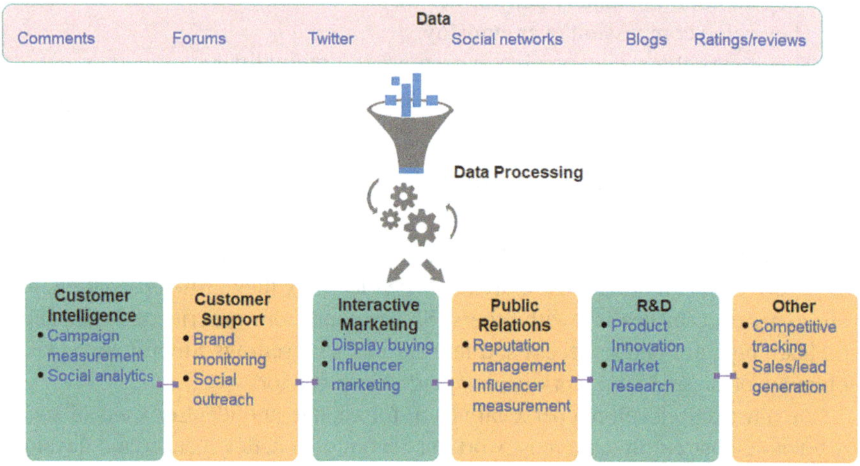


**Fig. 4.4** Social intelligence in business consists of 3 distinct activities: data collection, data processing, and insight delivery

## 4.5 Applications of Social Intelligence

Social intelligence is particularly useful in the various contexts. It represents a form of wisdom that encompasses social awareness, social understanding, and social interactions. It shows itself in the nursery, on the playground, at workplace, in barracks, in classroom, hotels, and factories. Here are a few ways SI is applied.

- *Management:* Social intelligence is important for businesses, managers, and management. It is a well-established fact that managers operate in a complex environment, with increasing levels of uncertainty further up hierarchical levels and competing demands from a complex social environment. The duty of a manager often involves directing the efforts of others in order to achieve desired results. Social intelligence is the manager's understanding of social interactions within organizations [10]. The social intelligence of managers can be vital in terms of the performance of employees. Social intelligence represents a manager's tacit knowledge, abilities, and skills to sense and understand the needs of customers and employees. The managers' levels of social intelligence affect employees' emotional labor and the emotional climate of the workplace. Managers with high social intelligence manage the emotional labor of their employees [11]. As shown in Fig. 4.4, social intelligence in business consists of three distinct activities: data collection, data processing, and insight delivery.
- *Leadership:* Leadership is social in nature. Since social intelligence is the ability to get along well with others and to get them to cooperate with you, it is needed in leadership. There is a noticeable difference between socially intelligent and socially unintelligent leaders in their performance. SI is essential for effective leadership; it helps teams work better together. Leaders can change if they are ready to put in the necessary effort. A socially intelligent leader will avoid behavior or activity that brings out negative emotions in others.
- *Workplace:* Social intelligences has indirect but great impact on workplace culture. To feel loved and happy at home and at work requires cultivating social intelligence. No one will be able to function successfully in a team without SI. People feel and react differently towards situations based on their cultural backgrounds. A high degree of social intelligence can also help you keep your job and make an impact on your workplace [12].
- *Classroom Discipline:* Teaching requires the careful handling of students' and teachers' own emotions. Social intelligence is a prerequisite for teachers because it helps teachers anticipate future behavior of students. For example, they need SI for classroom discipline. Classroom discipline management strategies are a set of interactions that assist teachers to influence students' behavior and teach them to act positively [13].
- *Social Computing*: The rapid development in technologies has changed the interaction between individuals. Human survival requires that new technologies be designed to support our adaptation to a changing environment. This in turn alters the structure and practices of social interaction. People are increasingly using

blog, forum, and bulletin boards to express their experiences. Social computing recreates facets of what it means to be human through technology [14].

- *Social Intelligence Design:* This aims to establish an intelligence information processing technology to support the process of sharing common knowledge within the society. Social intelligence design aims at building a system that is embedded in human society and supports interpersonal relationships. In the meantime, considerable emphasis is also placed in this field on the observation and analysis of social intelligence as it is manifested in human society. Social intelligence design research is interdisciplinary, borrowing from conceptual, scientific, and engineering viewpoint [15]. It supports community building by embedding human social intelligence into web applications.
- *Artificial Social Intelligence* (ASI): This is the application of machine intelligence techniques to social phenomena. ASI includes the computer simulations of societies or groups of organizations in which the individuals are modeled as intelligent actors. Traditionally, computer scientists working on AI have largely ignored sociology and social intelligence. Advances in computer science have brought technology to the point where several applications suggest that ASI could be of great significance to sociology, which is primarily concerned with the behavior of societies of complex, adaptive individuals. ASI has the potential to revolutionize sociology [16].
- *Social Neuroscience:* This branch of science is mainly concerned with how the brain drives social behavior and in turn how our social world influences our brain. It deals with the neural mechanisms that orchestrate our interactions and relationships. Social cognitive neuroscience is an interdisciplinary approach. From social neuroscience viewpoint, social intelligence comprises both cognitive and non-cognitive aptitudes [17].

Other areas of applications of SI include social science, social learning, social influence, marketplace metacognition, games, music education, moral instruction, and acting.

## 4.6 Assessing Social Intelligence

Social IQ is a measure of social intelligence, which has until recently been measured by various techniques. The most common techniques of assessment are tests, test situations, rating of social traits, and questionnaires. One way to measure social IQ is to use the basic IQ system, adapted for social skills. Measuring social intelligence involves identifying key interaction skills and then assessing them behaviorally. Several studies employ performance-based measures to assess various aspects of social intelligence. One of these was the George Washington Social Intelligence Test (GWSIT). The GWSIT is probably been the most widely used. It consists of a number of subtests, which can be combined to yield an aggregate score. This test came under immediate criticism for its relatively high correlation with abstract intelligence [18]. A Sociability Test for elementary-school children has been designed by Hsia. Other tests include the Dymond Rating Test, the Six Factor Test of Social Intelligence, Social Participation Scale, Tromso Social Intelligence Scale (TSIS), the SIPS questionnaire (Solution of Interpersonal Problem-oriented Situations), and a test of general mental ability (GMA).

## 4.7 Benefits

Social intelligence refers to the ability to successfully build relationships and navigate social environments. It is the essence of human relationships. As social beings, we need the cooperation of others. Social intelligence helps individuals build and maintain relationships. It is important in crisis situations.

Increasing a person's social intelligence will provide benefits throughout his professional and personal lives. It can help promote mutual respect, active listening, and lead to successful intercultural collaboration. Improving your social intelligence will make you become socially smarter. Working towards a strong social intelligence can lead to a richer life. It can also help you keep your job and thrive. It makes it easier to make new friends. When you understand what makes other people tick, you are displaying social intelligence.

Social intelligence is related to wisdom, values, and personality traits. It may be regarded as the major building block of building and maintaining social relationships. It facilitates quick processing of social information. Effective conflict resolution, successful negotiation, great customer service, increasing flexibility, efficient communication, enhancing relationships, and improving decision-making skills are all the main objectives of social intelligence. Socially intelligent people act wisely in maintaining human relations. They avoid using words that tend to intimidate, abuse, or threaten others. Teaching social intelligence is an indispensable luxury because it is the best resource to build and maintain a good work culture.

## 4.8 Challenges

Although the concept of social intelligence has been around for a long time, it faces some challenges. In spite of its intuitive appeal, social intelligence is difficult to measure. Social intelligence is a multidimensional construct that people use to solve their daily life problems. A major challenge is having the time to socialize and develop social skills.

Some critics argue that SI does not necessarily apply across all situations. It remains a difficult construct to operationalize. Tests of social intelligence are hard to devise. Social intelligence do not always have trackable and tangible results. Social intelligence may be used to harm others as typified by bullying behavior, aggression, workplace abuse, gossip, and rumor.

## 4.9 Conclusion

Social intelligence is an individual's competence to understand one's environment and react appropriately for socially successful conduct. It plays a crucial role in an individual's socialization process and professional development. It paves the way for cultural intelligence, social reform, and social activities that are intended to improve human well-being. There is still a great deal to be known about social intelligence. Science and technology will continue to advance our thinking about SI [19]. More information on social intelligence is available in the books in [20–28] and other books available on Amazon.com.

## References

1. R. Thakur, S. Sharma, R. Pathania, Social intelligence of adolescents: A study of Himachal Pradesh. Stud. Home Community Sci. **7**(3), 161–166 (2013)
2. M.N.O. Sadiku, S. Alam, S.M. Musa, Social intelligence: A primer. International Journal of Research-Granthaalayah **7**(9), 213–217 (2019)
3. Social intelligence, Wikipedia, the free encyclopedia https://en.wikipedia.org/wiki/Social_intelligence
4. A. Kaukiainen, The relationships between social intelligence, empathy, and three types of aggression. Aggress. Behav. **25**, 81–89 (1999) https://www.toolshero.com/communication-skills/social-intelligence/
5. R.E. Riggio, What is social intelligence? Why does it matter? https://www.psychologytoday.com/us/experts/ronald-e-riggio-phd
6. B. Wiest, 16 signs of a socially intelligent person, https://www.huffingtonpost.com/brianna-wiest/16-signs-of-a-socially-in_b_8855184.html
7. B. Wiest, 13 things socially intelligent leaders do differently, https://www.forbes.com/sites/briannawiest/2018/03/13/13-things-socially-intelligent-leaders-do-differently/#1778f434dbd2
8. A. Morin, C. Snyder, How to increase your social intelligence (2020), https://www.verywellmind.com/what-is-social-intelligence-4163839
9. Leading with character: Social intelligence, https://www.michaelleestallard.com/leading-with-character-social-intelligence#:~:text=Social%20intelligence%20is%20one%20of,leading%20is%20social%20in%20nature.&text=Given%20this%2C%20leaders%20should%20be,and%20feel%20as%20they%20do
10. A.J. Morris, Perceptions of social structure, others and self: The role of social intelligence in managerial effectiveness, Doctorial Dissertation, University of California, Irvine (1997)
11. Five ways to increase your social intelligence, http://www.socialigence.net/blog/5-ways-to-increase-your-social-intelligence/
12. 4 Reasons why social intelligence matters at work (2018), https://www.skillsportal.co.za/content/4-reasons-why-social-intelligence-matters-work
13. S.Y. Jeloudar et al., The influence of social intelligence of secondary school teachers on classroom discipline strategies. J. Psychol. **3**(1), 39–45 (2012)
14. S.P. Goggins, M. Gallagher, C. Amelung, Social intelligence in completely online groups toward social prosthetics: From log data analysis and transformation, in *IEEE International Conference on Social Computing/IEEE International Conference on Privacy, Security, Risk and Trust* (2010), pp. 500–507
15. T. Nishida, Social intelligence design for cultivating shared situated intelligence, in *Proceedings of the IEEE International Conference on Granular Computing* (2010), pp. 369–374

16. E.E. Brent et al., Artificial social intelligence. Annu. Rev. Sociol. **20**, 407–436 (1994)
17. D. Wawra, Social intelligence. Eur. J. Engl. Stud. **13**(2), 163–177 (2009)
18. J.F. Kihlstrom, N. Cantor, Social Intelligence, in *Handbook of Intelligence*, ed. by R. J. Sternberg, 2nd edn., (Cambridge University Press, Cambridge, UK, 2000), pp. 359–379
19. J. Veronesi, The lowdown on social intelligence. Home Health Care Manag. Pract. **21**(3), 203–204 (April 2009)
20. E. N. Goody (ed.), *Social Intelligence and Interaction: Expressions and Implications of the Social Bias in Human Intelligence* (Cambridge University Press, 1995)
21. D. Goleman, *Social Intelligence: The New Science of Human Relationships* (Hutchinson, 2006)
22. R. J. Sternberg (ed.), *Handbook of Intelligence*, 3rd edn. (Cambridge University Press, Cambridge, UK, 2011)
23. K. Albrecht, *Social Intelligence: The New Science of Success* (Pfeiffer, 2009)
24. S.J. Sampson et al., *Social Intelligence Skills for Correctional Supervisors/Managers* (HRD Press, 2006)
25. C.R. Epstein, A. Pawar, S.C. Simon, *Emergency Management and Social Intelligence: A Comprehensive all-Hazards Approach* (CRC Press, Boca Raton, FL, 2015)
26. T. Buzan, *The Power of Social Intelligence* (Thorsons, 2002)
27. R.S. Wyer, T.K. Srull, *Social Intelligence and Cognitive Assessments of Personality* (Psychology Press, 1989)
28. N.E. Snow, *Virtue as Social Intelligence: An Empirically Grounded Theory* (Routledge, 2010)

# Chapter 5
# Spiritual Intelligence

## Contents



Spiritual intelligence is the capacity to conduct our life in such a way that it reflects deep philosophical and metaphysical understanding of reality and of ourselves discovered through personal experience during systematic spiritual pursuit.

—Stanislav Grof

## 5.1 Introduction

Human intelligence has always been intriguing psychologist, educators, researchers, and practitioners. Intelligence refers to the ability to learn from experience and to acquire and retain knowledge. It is simply to use what you know in the right way with the right intention. It is largely believed to be inborn and difficult to alter. It can be measured by instruments called IQ tests.

The most widely used intelligence is rational or intellectual intelligence (IQ). When we talk about multiple intelligences, we often refer to physical, cognitive, emotional, moral, spiritual, musical, or spatial intelligence. One may be highly developed in one of the dimensions but not in others. Spiritual quotient (SQ) indicates parallels with IQ (Intelligence Quotient) and EQ (Emotional Quotient). It has been called the ultimate intelligence. As shown in Fig. 5.1, spiritual intelligence is


M. N. O. Sadiku, S. M. Musa, *A Primer on Multiple Intelligences*,
https://doi.org/10.1007/978-3-030-77584-1_5

related to other intelligences: physical intelligence (PQ), mental or intellectual intelligence (IQ), emotional intelligence (EQ). If cognitive intelligence is about thinking and emotional intelligence is about feeling, then spiritual intelligence is about being. SQ is uniquely human. It is something we are all born with. It is possible that some people are born with a higher potential for spiritual intelligence than others. Unlike IQ which computers have, EQ exists in higher mammals.

Spiritual intelligence is a phrase that has started to become familiar among educators and psychologists. The term implies awareness of our relationship to the divine. It is a unique ability to feel, understand, and act beyond selfish motives. It expands our capacity to understand fellow human beings at the deepest level. Attributes of spiritual intelligence include faith, humility, gratitude, self-control, modesty, morality, forgiveness, wisdom, integrity, honesty, compassion, and love. Spiritual intelligence enables us to see things as they are. It helps us to make wise choices and learn from our mistakes [1].

This chapter examines the concept of spiritual intelligence as a basic quality from which all other human abilities arise. It begins by describing what spiritual intelligence is all about. It provides the components and principles of SI. It addresses the difference between spirituality and spiritual intelligence. It presents how spiritual intelligence can be developed. It discusses how spiritual intelligence is applied in different areas. It highlights the benefits and challenges of spiritual intelligence. It adds some quotations on spiritual intelligence. It concludes with some comments.

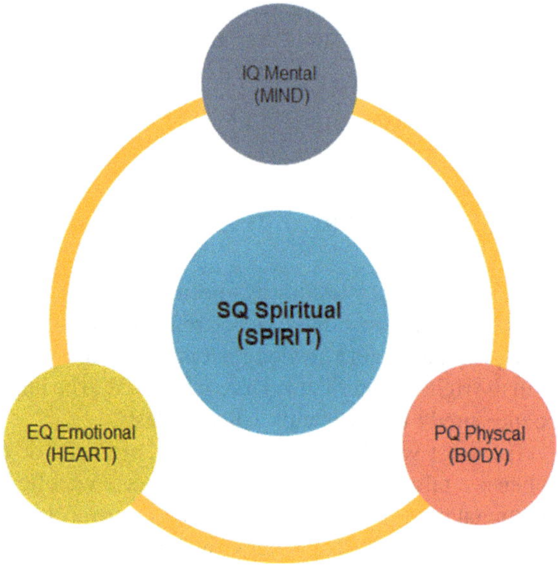


**Fig. 5.1** How spiritual intelligence is related to other intelligences

## 5.2 What Is Spiritual Intelligence?

The concept of spiritual intelligence (SQ) was first proposed by Howard Gardner in 1993 in his theory of "multiple intelligences." This theory suggests that intelligence is best understood as a series of related but separate cognitive processes and abilities. Different kinds of intelligence are associated with different areas of the brain.

Spiritual intelligence is concerned with integrating the inner life of mind and spirit with the outer life of work in the world. It is the ability to behave with compassion and wisdom while maintaining peace regardless of the circumstances. It is about how we behave, act, and make decisions on daily basis as we interact with difficult people and situations. It is essential for both personal and professional development. Every person has their innate spiritual capacities that give meaning and purpose to life.

Spiritual intelligence varies with individuals. Signs of high spiritual intelligence include an ability to think out of the box, love, compassion, and humility. SQ includes accessing wisdom, beauty, deeper meaning, higher purpose, goodness, truth, and exercising intuition, compassion, and other attributes.

Robert Emmons claimed that spiritual intelligence consists of several components. He proposed five components of spiritual intelligence [2, 3]:

1. The capacity to transcend the physical and material.
2. The ability to experience heightened states of consciousness.
3. The ability to sanctify everyday experience.
4. The ability to utilize spiritual resources to solve problems.
5. The capacity to be virtuous.

In addition to these components, Emmons provided the following seven major themes of spiritual intelligence [4]:

1. *Consciousness*: Developed refined awareness and self-knowledge
2. *Grace:* Living in alignment with the sacred manifesting love for and trust in life
3. *Meaning:* Experiencing significance in daily activities through a sense of purpose and a call for service, including in the face of pain and suffering
4. *Transcendence*: Going beyond the separate egoic self into an interconnected wholeness
5. *Truth:* Living in open acceptance, curiosity, and love for all creation (all that is)
6. *Peaceful surrender to Self* (True, God, Absolute, true nature)
7. *Inner-Directedness:* Inner-freedom aligned in responsible wise action.

Figure 5.2 shows various components of spiritual intelligence.

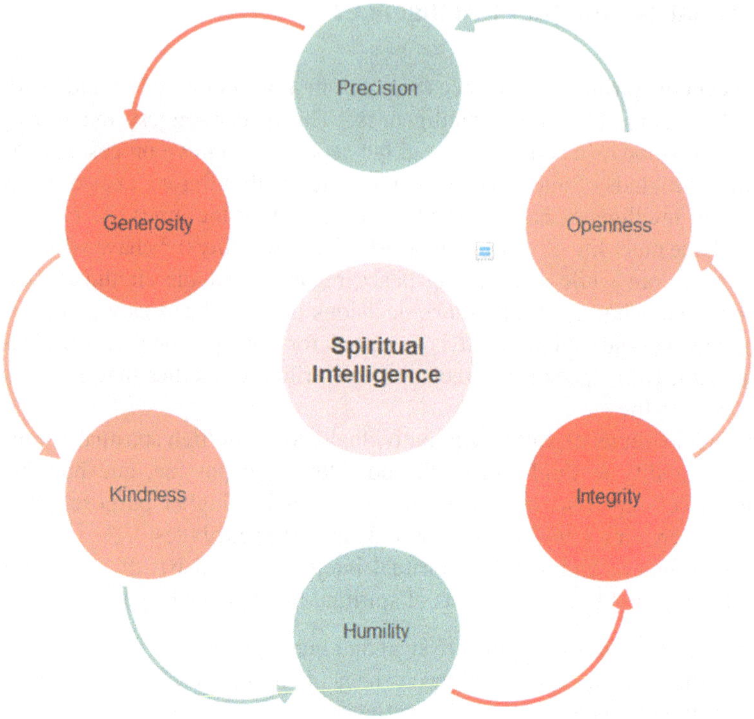


**Fig. 5.2** Components of spiritual intelligence

## 5.3 Principles of Spiritual Intelligence

One way to learn about spiritual intelligence is through its principles. Danah Zohar defined 12 principles underlying spiritual intelligence as follows [5]:

1. *Self-awareness*: Knowing what I believe in and value, and what deeply motivates me.
2. *Spontaneity:* Living in and being responsive to the moment.
3. *Being vision- and value-led:* Acting from principles and deep beliefs, and living accordingly.
4. *Holism*: Seeing larger patterns, relationships, and connections; having a sense of belonging.
5. *Compassion:* Having the quality of "feeling-with" and deep empathy.
6. *Celebration of diversity*: Valuing other people for their differences, not despite them.
7. *Field independence*: Standing against the crowd and having one's own convictions.
8. *Humility:* Having the sense of being a player in a larger drama, of one's true place in the world.

9. *Tendency to ask fundamental "Why?" questions:* Needing to understand things and get to the bottom of them.
10. *Ability to reframe*: Standing back from a situation or problem and seeing the bigger picture or wider context.
11. *Positive use of adversity*: Learning and growing from mistakes, setbacks, and suffering.
12. *Sense of vocation:* Feeling called upon to serve, to give something back.

These principles can help us to measure our spiritual intelligence. They will help us grow in our spiritual development and cultivate empathy and compassion for others. Living by the principles can help us lead richer lives and make an impact in the world.

## 5.4 Spirituality and Spirirual Intelligence

Any discussion of spiritual intelligence is incomplete without addressing spirituality.

World religions (such as Christianity, Islam, and Judaism) agree that humans consist of body, mind, soul, and spirit. A growing body of knowledge shows that spirituality can be "good medicine" for the body and mind. Therefore, spirituality and spiritual intelligence must be recognized as an integral and important part of the human condition. Religion, spirituality, and spiritual intelligence are different, yet connected.

- *Religion:* A religion is an institutionalized body of believers who accept a common set of beliefs, practices, and rituals. It is based on dogma and what you have to believe. It incorporates spiritual consciousness into a system of belief, worship, morality, and social responsibility. Religion has been defined as beliefs or practices that express spirituality, while spirituality is the intrapersonal experience of the divine. Religious life may consist in church, synagogue or mosque attendance and observance of rituals such as baptisms, weddings, and funerals.
- *Spirituality:* Spirituality begins with deep connection with God and refers to an aspect of reality that is more than physical, emotional or intellectual. It is often viewed as the innate human need to connect with something larger than ourselves. It is an individual's subjective beliefs and experiences about a power greater than oneself. Spirituality deals with our search for meaning and purpose. It is a way of life; it involves intellect, will, and emotion. Our personal sense of spirituality guides our choices and enables us to cope with life difficulties. Spiritual people believe in a God or higher source of life. Some have spiritual experience involving direct communion with a higher power. They connect to the divine through prayer, meditation, contemplation, and journaling [6]. Spirituality refers to a transcendence that is beyond the material realm. The spiritual realm is hidden because it cannot be detected by our intellect or by scientific means. Spirituality is a personal experience and is not one size fits all. This implies that

every employee in your organization can have different beliefs, needs, and sensitivities [7].

- *Spiritual intelligence*: This is the ability to apply and embody spiritual resources to enhance daily functioning and well-being. It can be developed by a variety of practices and training. Spiritual intelligence is our capacity for developing meaning, vision, and value. It frees you from neediness and clinginess in your personal relationships. Spirituality is to "know" who you are, while spiritual intelligence is to "realize" who you are and to live life in that awareness. Like spirituality, spiritual intelligence is considered multidimensional.

## 5.5 Developing Spiritual Intelligence

Developing our spiritual intelligence means moving from ego-driven behaviors to more spiritual behaviors and our actions become purpose and value driven. Our search for meaning and purpose in life is the primary motivation. When this need is not met, our productivity suffers. The price is paid in education, where the spiritual intelligence of children is not developed, and in the workplace [8]. Our spiritual intelligence cannot be separated from our being and should be nurtured as we grow. It can be developed with practice. Its attributes include faith, humility, gratitude, and ability to love and forgive [9].

Spiritual intelligence reflects the mental flexibility to make adjustments and confront life problems. It increases with age. It is not related to a particular religion or creed. Improving spiritual intelligence may help one become more stable and adaptable.

Ways to develop spiritual intelligence include [10, 11]:

1. Life purpose. What decisions and actions can you take that are in line with your purpose?
2. Personal values. Values shape our beliefs. Make a list of your ten most important personal values.
3. Life goals. Think about your goals and use visualization to see your goals, desires, and wants fulfilled.
4. Take responsibility for your goals, desires, and wants.
5. Develop a sense of community by letting more people into your life.
6. Understand other people's spiritual beliefs. Pay attention to the differences in your own worldview.
7. Make a commitment to spiritual growth. It is important to commit time to a regular practice, such as meditation, yoga, or journaling.
8. Read and meditate. Immerse yourself in great literature or great music.
9. Find it in the way you communicate with nature.
10. When chance knocks at your door, let it in and take advantage of coincidences.

11. Let love as our guide. Life is all about the decisions we make and we cannot go wrong when we focus on love and compassion.

## 5.6 Applications of Spiritual Intelligence

Spiritual intelligence is an emerging concept that refers to the ability to access higher meanings, values, abiding purposes, and unconscious aspects of the self. It addresses and solves many life problems. Individuals who are rich spiritually can best express their gifts, intellect, and vision in the world. Scholars have been studying the role of spirituality in different contexts, such as family life, leadership, healthcare, management consulting, business, the workplace, and education.

- *Family Life*: Family is the generator of manpower. Instability in family life puts individuals' mental and physical health in danger. Marital relationship is the central part of the family's spiritual and social life. Being spiritually intelligent in a family environment makes one to relate maturely and embrace others. Children learn to appreciate their unique talents and embrace the world with passion. Education has an impact on spirituality. Instability in married life puts individuals' mental and physical health in danger. Parents with a higher education have a higher spiritual intelligence [12]. Parents use their spiritual intelligence to discover what is important in life and help their children realize their potential for goodness. Life satisfaction is a subjective assessment of the quality of one's of spiritual intelligence. It is an overall assessment of feelings which may have negative or positive impact on family members.
- *Workplace:* Spiritual intelligence is the commitment to righteousness and wisdom at the workplace, which is constantly changing. It can be used as a factor of influence in work performance, which affect personal skills, hard work, and productivity. By developing our spiritual intelligence, we can develop skills such as intrapersonal and interpersonal relations, problem-solving, goal attainment, motivation, commitment, responsibility, team spirit, stress management, time management, necessary for work and be successful at workplace [13].
- *Business:* Today, businesses are facing intense scrutiny into how they operate. Economic pressures, technology, global interconnectivity, and demographic changes in the work force are changing the complexity of the business environment. Spiritual intelligence can be used to influence in work performance, which is important to the company in order to achieve company's goal. A growing number of companies such as Chick-Fil-A, Maxwell, Locke, Taco Bell, and Pizza Hut are now using spiritual lessons in their management and leadership training.
- *Leadership*: For leadership, the intelligences we deal with the most are the physical, emotional, social, cognitive, and spiritual lines. What makes a great leader is the ability to provide a clear sense of direction. Leaders in government, education, sports, social work, and business achieve their goals by motivating their followers, instilling trust, and promoting positive relationships. Spiritual

intelligence is essential for authentic leadership. It offers the potential of a compass through which a leader's leadership ability can be assessed. It is based on values, not on profits. It is a global intelligence exemplified by great leaders such as Jesus, Mohammed, Mother Teresa, Gandhi, Martin Luther King, and Nelson Mandela. We admire these as spiritual leaders who moved people emotionally (inspiration, feeling, or change), physically (to action), and mentally (to new ideas). Organizational leaders use spiritual intelligence to lead their organizations effectively, enhance organizational productivity, influence organizational trust, foster organizational commitment, and promote organizational culture. The family, community, and global business leaders of the future are those who cultivate the skills of spiritual intelligence. Leaders and employees with a higher level of spiritual intelligence often have a higher level of work satisfaction [14]. As shown in Fig. 5.3, spiritual intelligence manifests itself in leadership as love, giving, excellence, courage, and so on.

- *Healthcare*: Spiritual care is a growing topic of interest to healthcare practitioners. Since nursing is a discipline that deals with people, the inclusion of spirituality in care is vital. Nurses require special skills such as self-awareness, communication, trust, and hope in order to provide spiritual care. It has been

**Fig. 5.3** Spiritual intelligence in leadership

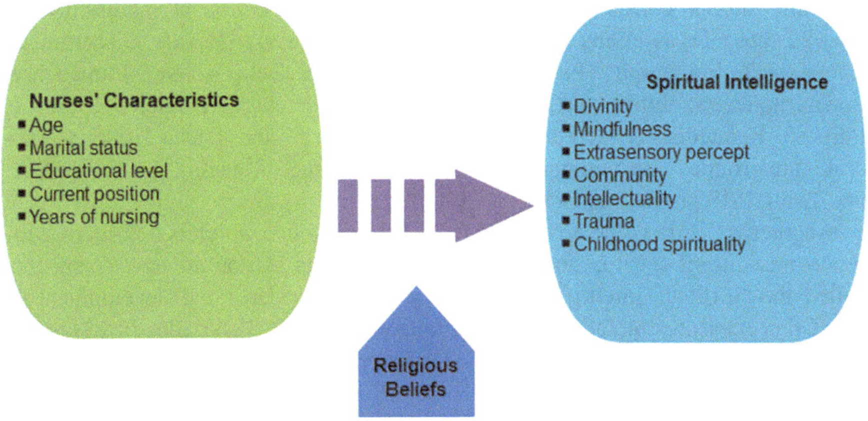


**Fig. 5.4** Spiritual intelligence in a nurse

observed that nurses who have more spiritual intelligence are happier than those who have lower spiritual intelligence. Figure 5.4 displays the spiritual intelligence of a nurse. Spirituality affects health habits and behaviors. It is considered as an important aspect of health in people with chronic conditions [15]. Improving spiritual intelligence among the elderly help them attain greater stability and maintain a better quality of life.

- *Education*: Education is a systematic process through which a person acquires knowledge. It makes an individual civilized, refined, and cultured. The ultimate goal of education is all round holistic development of the student. Spiritual intelligence plays an important role in education. It may be regarded as the ability in every human being for solving problems that are related to life, values, and education. Students' spirituality helps them to make good choices and cope with peer difficulties. Teachers need to realize that all students have spiritual intelligence, but differ in the level. This intelligence is not fixed but can be enhanced by educational opportunities [16]. Educating for spiritual development provides opportunities for students to develop their spiritual intelligence and discover what is important in life. It fills the vacuum in people's lives.

Spiritual intelligence can also be applied in spiritual care, teamwork, English writing, engineering education, neurosciences, and mental health.

## 5.7 Assessing Spiritual Intelligence

Spiritual intelligence of an individual can be developed and measured. Several self-report measures have been developed to assess spiritual intelligence. Perhaps the most popular measure of spiritual intelligence is the Integrated Spiritual Intelligence Scale (ISIS), which is an 83-item long form; ISIS has shown satisfactory factor structure, internal consistency, test-retest reliability, and construct validity. It contains 22 subscales assessing separate SI capabilities related to Beauty, Discernment, Egolessness, Equanimity, Freedom, Gratitude, Higher-self, Holism, Immanence, Inner-wholeness, Intuition, Joy, Mindfulness, Openness, Practice, Presence, Purpose, Relatedness, Sacredness, Service, Synthesis, and Trust. These 22 subscales are grouped into 5 domains: Consciousness, Grace, Meaning, Transcendence, and Truth. ISIS predicts satisfaction with life [17].

Another measure is the Scale for Spiritual Intelligence which is a 20-item, self-report measure of spiritual intelligence in adolescents. An online assessment tool called the Spiritual Intelligence Inventory (SQi) measures the development of SQ. Other measures include Spiritual Intelligence Scale (SIS) defined by Nasel in 2004, the Emotional Intelligence Scale (WLEIS) proposed by Wong and Law, the Spiritual Intelligence Self Report Inventory (SISRI) developed by David King in 2008, Spiritual Intelligence Questionnaire made in 2009 by Abdollah Zadeh, the Intrinsic Spirituality Scale (ISS), the 62-item Chinese Spiritual Intelligence Scale (CSIS), Islamic spiritual intelligence (ISI), and the Beliefs about Consciousness and Reality Questionnaire (BACARQ) created by Baruss and Moore in 1992.

## 5.8 Benefits

Spiritual intelligence is the ultimate intelligence reflecting values associated with mental adaptation capacities and leading to non-materialistic life view. It helps us to grow spiritually and live with no pressure, concern, fear of change, and anxiety. It is associated with elderly life satisfaction [18]. Mature, flexible, skillful leadership is connected with the skills of spiritual intelligence. Spiritual beliefs and practices are linked with physical health, marital satisfaction, and stability. Spiritual intelligence enables us to persist even under the biggest pressures. It gives meaning and purpose in life. A high spiritual intelligence is the best predictor of happiness, serenity, good self-esteem, wisdom, peace, love, and harmonious relationships. It will not only benefit individuals, it will also benefit their families, employers, and society. Spiritual intelligence has been found to have a positive correlation with mental well-being or mental health. It is crucial in facing challenges and facilitating daily problem solving.

Smith described ten skills originating from spiritual intelligence as follows [19]:

1. *Spiritual experiences:* Certain religious activities and experiences
2. *Stress coping*: Using religion and faith to solve the life problems
3. *Purpose:* Having a specific purpose in life by considering religious issues
4. *Worship place*: Willingness to religious places and religious leadership
5. *Living out*: Evading from the tenets and stereotypes in life
6. *Centrality*: Impact of religion on behavior and performance (such as eating, drinking, wearing)

7. *Religious practices:* Compliance with laws and commands of religion and faith in life.
8. *Prayer and religious practices*
9. *Tolerance:* Tolerance of other religions and correct treatment based on beliefs, principles, and logic
10. *Religious concepts*: Belief in religious basic concepts (such as the Creator of the world, soul, life after death)

## 5.9 Challenges

Some say that intelligence is not really the best word to use when referring to spirituality. They consider wisdom as more appropriate. Like other intelligences, spiritual intelligence is difficult to empirically measure or quantify. All efforts to standardize models of spiritual intelligence over diverse settings of complex religious and cultural compositions should avoid secularizing them. Spirituality is a subjective experience which may mean different things to different people. Logical reason alone is inadequate in explaining the spiritual experience of individuals. Since spiritual intelligence is relatively new as a research discipline, a single standard for determining spiritual intelligence does not exist.

## 5.10 Quotes on Spiritual Intelligence

Quotes on spiritual intelligence include the following [20].

> Intelligence is the ability to understand the problem and find a solution to that problem.—Awdhesh Singh

> Spiritual intelligence is the tenth intelligence that integrates all other intelligences.—Dorothy Sisk

> All great leaders are known to have faith. They first develop faith in themselves, which means they connect their body, mind, and intellect with their soul. They also connect their soul with the Universal Soul where all souls of the world are connected with each other. Leadership is thus a spiritual quality that comes from faith and trust.—Awdhesh Singh

> Spiritual intelligence is the capacity to conduct our life in such a way that it reflects deep philosophical and metaphysical understanding of reality and of ourselves discovered through personal experience during systematic spiritual pursuit.—Stanislav Grof

> Spiritual intelligence represents our drive for meaning and connection with the infinite.—Stephen Covey

> When emotional intelligence merges with spiritual intelligence, human nature is transformed.—Deepak Chopra

The harmony of natural law reveals an Intelligence of such superiority that, compared with it, all the systematic thinking and acting of human beings is an utterly insignificant reflection.—Albert Einstein

Man is a spiritual intelligence, who has taken flesh with the object of gaining experience in worlds below the spiritual, in order that he may be able to master and to rule them, and in later ages take his place in the creative and directing hierarchies of the universe.—Annie Besant

Never place a time limit on the healing. That is dictating to the spiritual intelligence that does its perfect work when it is not impeded by your questions and false beliefs.—Raymond Charles Barker

Through persistent, effective, and diligent work, a person can accumulate knowledge in the form of facts, data, information, and experience. Intelligence, however, can only be gained through obedience. Thus, knowledge is a prerequisite to and foundation for true spiritual intelligence.—David A. Bednar

Each visible planet is the embodiment of a great and exalted spiritual intelligence. Who is the minister of God in that department of His Kingdom, endeavoring to carry out His Will, the latter having in view the ultimate highest good, regardless of temporary ill." Max Heindel

Being spiritual is a far more intelligent way to exist. It means being in tune with the intelligence of the Creator.—Jaggi Vasudev

I am attracted to intelligence, a witty sense of humor, an adventurous outlook on life and spiritual awareness about one's self and the world.—Tanit Phoenix

Cultures are never merely intellectual constructs. They take form through the collective intelligence and memory, through a commonly held psychology and emotions, through spiritual and artistic communion.—Tariq Ramadan

It is an accepted commonplace in psychology that the spiritual level of people acting as a crowd is far lower than the mean of each individual's intelligence or morality.—Christian Lous Lange

I consider myself a spiritual atheist. I certainly believe there are forces bigger than ourselves, and that we should be searching, individually, for meaning in our lives. But I don't believe there's a supreme being, an intelligence that created everything.—Alan Lightman

Husbands, recognize your wife's intelligence and her ability to counsel with you as a real partner regarding family plans, family activities, and family budgeting. Don't be stingy with your time or with your means. Give her the opportunity to grow intellectually, emotionally, and socially as well as spiritually.—Ezra Taft Benson

There is no greater power in heaven or on earth than pure, unconditional love. The nature of the God force, the unseen intelligence in all things, which causes the material world and is the center of both the spiritual and physical plane, is best described as pure, unconditional love.—Wayne Dyer

> Scientology is used to increase spiritual freedom, intelligence, ability and to produce immortality.—L. Ron Hubbard

> Having the Creator in my life, having the spiritual connection to the divine intelligence that created me.—John Assaraf

## 5.11 Conclusion

Science and spirituality are beginning to take a closer look at one another and interact. Spiritual intelligence is the ability to access higher meanings and values and allow these meanings and values live richer and creative life. It is a way to make meaning in life. It is necessary in making choices that contribute to our well-being. It helps to discover the purpose of life and make sense of living. People are different and they are at different levels of spiritual intelligence. It is highly recommended that educational institutions (from elementary to high institutions) should expand their studies to include spiritual intelligence. More information about spiritual intelligence can be found in the books in [21–37] and two related journals: *Religions* and *International Journal on Spirituality* and *Organization Leadership.*

## References

1. M.N.O. Sadiku, A. Omotoso, S.M. Musa, Toward better understanding of spiritual intelligence. Int. J. Trend Scient. Res. Develop. **4**(4), 1113–1115 (2020)
2. Spiritual intelligence, Wikipedia, the free encyclopedia, https://en.wikipedia.org/wiki/Spiritual_intelligence
3. R.A. Emmons, Spirituality and intelligence: problems and prospects. Int. J. Psychol. Relig. **10**(1), 57–65 (2000)
4. P.S. Srivastava, Spiritual intelligence: an overview. Int. J. Multidiscip. Res. Develop. **3**(3), 224–227 (2016)
5. 12 Principles of SQ, https://www.danahzohar.com/twelveprinciples/
6. C. Kilcup, Secret wisdom: spiritual intelligence in adolescents. Gift. Educ. Int. **32**(3), 242–257 (2016)
7. How to achieve spiritual intelligence (2009), https://www.wellbeing.com.au/mind-spirit/spirituality/spiritual-intelligence.html
8. D.A. Sisk, Spiritual intelligence: developing higher consciousness revisited. Gift. Educ. Int. **32**(3), 194–208 (2016)
9. N. Ronel, The experience of spiritual intelligence. J. Transpers. Psychol. **40**(1), 100–119 (2008)
10. C. Wigglesworth, Developing spiritual intelligence, https://www.alchemyformanagers.co.uk/topics/KdveUGzCAz6JwS5S.html
11. D.A. Sisk, Spiritual intelligence: the tenth intelligence that integrates all other intelligences. Gift. Educ. Int. **16**(3), 208–212 (2002)
12. G. Mohammadyari, Relationship between parent's spiritual intelligence, level of education and children's mental health. Procedia. Soc. Behav. Sci. **69**, 2114–2118 (2012)
13. C. Joseph, S.S. Lakshmi, Spiritual intelligence at work. IUP J. Soft Skills **5**(4), 21–30 (2011)

14. M. Korazija, S.S. Zizek, D. Mumel, The relationship between spiritual intelligence and work satisfaction among leaders and employees. Nase Gospodar. Our Econ. **62**(2), 51–60 (2016)
15. M. Rahmanian et al., The relationship between spiritual intelligence with self-efficacy in adolescents suffering type 1 diabetes. J. Educ. Health Promot. **7**, 100 (2018)
16. G. DeBlasio, The effect of spiritual intelligence in the classroom: god only knows. Int. J. Child. Spiritual. **16**(2), 143–150 (2011)
17. Y. Amram, D.C. Dryer, The integrated spiritual intelligence scale (ISIS): development and preliminary validation, in *Proceedings of the 116th Annual Conference of the American Psychological Association*, Boston (2008)
18. M. Sahebalzamani et al., The relationship between spiritual intelligence with psychological Well-being and purpose in life of nurses. Iran. J. Nurs. Midwif. Res. **18**(1), 38–41 (2013)
19. A. Arbabisarjou et al., The relationship between spiritual intelligence and emotional intelligence among students at Isfahan University of medical sciences with a concentration on improvement of social relations. Int. J. Med. Res. Health Sci. **5**(9S), 596–603 (2016)
20. Quotes about spiritual intelligence, https://www.goodreads.com/quotes/tag/spiritual-intelligence#:~:text=%E2%80%9CSpiritual%20Intelligence%20is%20the%20Intuitive,desired%20objectives%20of%20the%20world.%E2%80%9D&text=%E2%80%9CSpiritual%20intelligence%20is%20like%20a,greed%2C%20arrogance%20and%20tyranny.%E2%80%9D&text=%E2%80%9CWhen%20you%20know%20the%20word,it%20suppose%20to%20change%20you!
21. A. Singh, *Practising Spiritual Intelligence: For Innovation, Leadership and Happiness* (Wisdom Tree Publishers, 2015)
22. B. Draper, *Spiritual Intelligence: A New Way of Being* (Lion Hudson, 2009)
23. A. Rossiter, *Developing Spiritual Intelligence: The Power of You* (O Books, 2006)
24. R.A. Bowell, *The 7 Steps of Spiritual Intelligence: The Practical Pursuit of Purpose, Success and Happiness* (Nicholas Brealey Publishing, 2005)
25. D.A. Sisk, E.P. Torrance, *Spiritual Intelligence: Developing Higher Consciousness* (Creative Education Foundation, Buffalo, 2001)
26. A.E. Nelson, *Spiritual Intelligence: Discover Your SQ. Deepen Your Faith* (Baker Books, Grand Rapids, MI, 2010)
27. W.R. McCollum, *How to Use Emotional, Cultural, and Spiritual Intelligence to Mentor Doctoral Learners: Best Practices and Tools to Help Mentors and Doctoral Learners Navigate the Dissertation Process* (McCollum Enterprises, Fort Washington, MD, 2015)
28. D. Zohar, I. Marshall, *SQ-Spiritual Intelligence, the Ultimate Intelligence* (Bloomsbury Publishing, London, 2000)
29. K. Vallotton, *Basic Training for Spiritual Intelligence: Develop the Art of Thinking Like God* (Chosen Books, 2021)
30. C. Wigglesworth, J. Mackey, *SQ21: The Twenty-One Skills of Spiritual Intelligence* (Select Books, 2015)
31. B. Steinhauer, *In Search of Spiritual Intelligence* (Fisher King Publishing, 2020)
32. J. Phillips, *No Ordinary Time: The Rise of Spiritual Intelligence and Evolutionary Creativity* (Livingkindness Foundation, 2011)
33. M. Brewer, *What's Your Spiritual Quotient?: How to Respond to Life's Toughest Issues with Spiritual Intelligence* (Destiny Image Publishers, 2008)
34. R. Wolman, *Thinking With Your Soul: Spiritual Intelligence and Why It Matters* (Harmony, New York, 2001)
35. B. Drape, *Spiritual Intelligence: A New Way of Being* (Lion Hudson, 2009)
36. K.D. Noble, *Riding the Windhorse: Spiritual Intelligence and the Growth of the Self* (Hampton Press, 2001)
37. F.J. MacHovec, *Spiritual Intelligence, Behavioral Sciences, and the Humanities* (Edwin Mellen Press, 2002)

# Chapter 6
# Emotional Intelligence

**Contents**



The essential difference between emotion and reason is that emotion leads to action while reason leads to conclusions.

—Donald Calne

## 6.1 Introduction

Humans are predominantly emotional creatures. Emotion forms the basis for our entity and gives meaning to our lives. Emotions are everywhere. Emotions constitute one of the three legs of the tripod that determine human interactions: emotion, cognition, and motivation. They appear in many forms—facial expressions, tone of voice, body language, and even works of art. Distinct state emotions include anger, disgust, fear, anxiety, sadness, happiness, joy, relaxation, and desire. Emotions matter. Our emotions influence every aspect of our lives. Emotions help us in making decisions, understand the world. Emotions can help or hurt you. Displaying and recognizing emotions is an important aspect of our daily social interaction. Emotion has become a mainstream topic in music, sound, images, video, and text [1]. Roseman's theory uniquely defines 17 emotions by combining the seven appraisal




M. N. O. Sadiku, S. M. Musa, *A Primer on Multiple Intelligences*, https://doi.org/10.1007/978-3-030-77584-1_6

dimensions as shown in Fig. 6.1. He outlined nature and degree of emotions, their impact, and possible cause.

Intelligence, on the other hand, refers to the ability to reason validly about a domain of information. It is capacity of the individual to act purposefully, to think rationally, and to deal effectively with his environment. Traditionally, intelligence has been linked with performance in IQ tests and people used to view intelligence and emotions as opposites. Emotional intelligence (EI) is a type of social intelligence that combines emotion and intelligence by viewing emotions as useful sources of information that help us to navigate our social environment. The emotional intelligence quotient (EQ) is a flexible group of skills, which can be learned and acquired. Someone has well said that [2]:

$$\mathrm{IQ} + \mathrm{EQ} = \mathrm{SUCCESS}$$

As shown in Fig. 6.2, emotional intelligence is part of a whole person. Social intelligence (SI), or street smarts, is the ability to understand and manage others to act wisely in their social interactions. It is about understanding your environment and having a positive influence on your social interactions. It is the key to life and career success [3].

This chapter provides an introduction on emotional intelligence. It begins by addressing what emotional intelligence is all about. It then presents the three major models of EI. It covers characteristics of emotional intelligence and emotional intelligent people. It provides some tips on how to develop EI. It covers some common

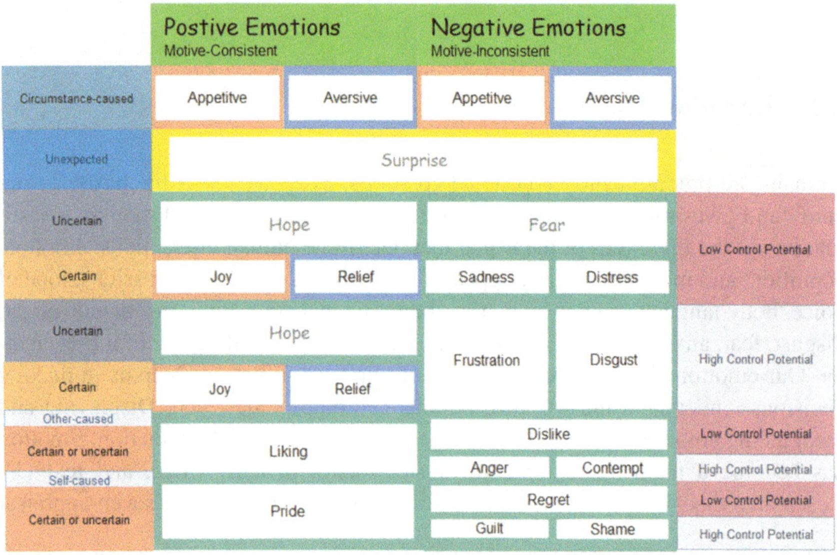


**Fig. 6.1** Roseman's structure of the emotions

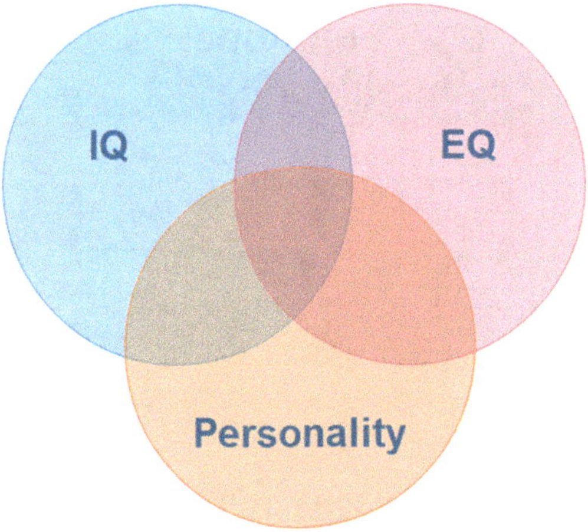


**Fig. 6.2** Emotional intelligence is part of a whole person

applications of EI. It highlights some benefits and challenges of EI. It adds some popular quotes on EI. The last chapter concludes with comments.

## 6.2 Concept of Emotional Intelligence

Charles Darwin was the first to recognize the value of emotions. The term "emotional intelligence" was coined by Peter Salovey of Yale University and John Mayer of University of New Hampshire in 1990. Emotional Intelligence (EI), or Emotional Intelligence Quotient (EQ), refers a person's ability to recognize, understand, control, and manage their emotions and as well as those of others. EI has been recognized for its correlation to success in work and personal life, regardless of industry. It has been the cornerstone of empowered energetic corporate cultures for years. It plays an important role for employees in any business of any size. It allows one to think more creatively and apply emotions to solve problems [4].

Emotional intelligence (or EQ) is often regarded as people skills—how we express ourselves, relate to others, and maintain social relationships. An individual who is in tune with his or her own emotions is much more likely to be able to understand and empathize with the emotions that impact the attitudes and behaviors of others. Intelligence quotient (IQ) by itself is only a partial indicator of success. Decades of research shows that emotional intelligence is the critical factor that sets star performers apart from others.

Just like traditional IQ, every person possesses a level of EI. Figure 6.3 shows low or high emotional intelligence. Emotions, like anger, happiness, and fear influence how people think, make decisions, and perform tasks. People have different personalities, traits, and values, and different ways of expressing their emotions. We unconsciously use emotional intelligence in our daily lives. Emotional intelligence promises to help people achieve success in life [5].

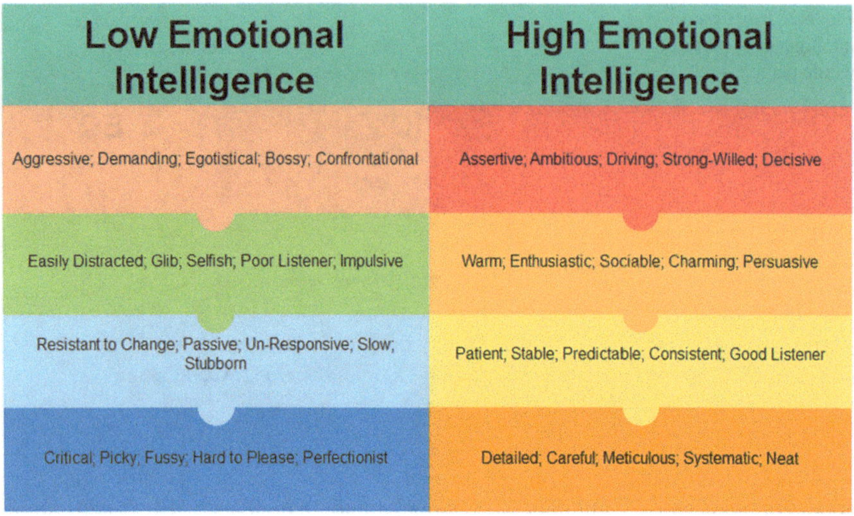


**Fig. 6.3** Low/high emotional intelligence

## 6.3 Models of Emotional Intelligence

Five models have been proposed as a means of better defining EI: the Ability Model, the Trait Model, the Mixed Model, the Bar-On model, and the Genos model [6]. We will consider only three of these. The models are presented in historical order [7].

***Model 1: Mayer and Salovey's Four-Branch Model*** John Mayer and Peter Salovey first introduced the term "emotional intelligence" in 1990 and have continued to conduct research on EI. Their model of EI includes four types of abilities and it is called ability model:

1. Perceiving emotions—the ability to identify and interpret emotions.
2. Using emotions—the ability to use emotions to facilitate various cognitive activities, such as thinking and problem-solving.
3. Understanding emotions—the ability to comprehend emotion language and to appreciate complicated relationships among emotions.
4. Managing emotions—the ability to regulate emotions in both ourselves and in others.

This model views emotions as useful sources of information that help one to make sense of and navigate the social environment. It assumes that individuals vary in their ability to process information [8].

***Model 2: Goleman's EI Model*** This model is the most popular. Goleman's model focuses on EI as a wide array of competencies and skills that drive leadership performance. Emotional competencies are not innate traits, but rather learned capabilities that can be improved upon. His model outlines five main EI constructs: (1)

Self-awareness, (2) Self-regulation, (3) Motivation, (4) Empathy, (5) Social skills. These will be discussed further in the next section.

***Model 3: Bar-On's Model*** Reuven Bar-On is the director of the Institute of Applied Intelligences in Denmark. In 1996, he explained that emotional intelligence reflects our ability to deal successfully with other people and with our feelings. In 2006, he developed the first measuring tool of emotional intelligence that was used as the term—Emotional Quotient. His model is a mixed model which has two components: the conceptual model and the psychometric model. The model suggests that EI competencies and skills contribute to how we understand ourselves and others.

## 6.4 Characteristics of Emotional Intelligence

The five elements that characterize emotional intelligence were suggested by Daniel Goleman, an American psychologist who helped to popularize emotional intelligence [9].

1. *Self-Awareness*: This is the ability to know your emotions, strengths, weaknesses, drives, and goals. This is the foundation upon what most other EI elements are built. Managers should help their workers improve self-awareness and more effectively manage their stress.
2. *Self-Regulation:* This is the ability to stay calm when emotions are running high. It involves thinking before doing and expressing your feelings maturely with restraint. Through greater self-awareness, leaders can regulate or control their emotional responses to tense situations.
3. *Motivation:* This constitutes emotional factors that help one reach their goals. Leaders hold themselves to a higher standard and provide guidance. High EQ people are self-motivated, pursuing personal goals for self-development and self-gratification.
4. *Empathy:* Empathy (understanding others, listening well, and reading nonverbal cues) is an important aspect of a manager. It is the ability to understand others' situation and have compassion on them. Unlike sympathy, empathy involves sharing the emotional experience of others.
5. *Social Skills*: These diverse skills are used to induce desirable responses in others. They often refer to how a leader handles and influences other people and their emotions. Emotionally intelligent people easily build trust and earn respect from others.

Thus, emotional intelligence is a combination of self-awareness, self-regulation, motivation, empathy, and social skills. Notice that all of these are *abilities*. These elements are illustrated in Fig. 6.4. Working on these five elements (self-awareness, self-regulation, motivation, empathy, and social skills) will increase your emotional intelligence or help you identify areas of improvement. Individuals that adopt these characteristics have a greater chance of being successful than individuals that do not.

In addition to the characteristics of emotional intelligence, the characteristics of emotionally intelligent individuals include the following [10].

1. They stay positive and manage difficult situations successfully.
2. They make thoughtful decisions.
3. They keep their emotions under control and discuss sensitive issues maturely and thoughtfully.
4. They are empathetic to their workers.
5. They have a strong positive influence on their team.
6. They are calm under pressure.
7. They know how to earn the respect of clients and employees.
8. They listen more often than they speak.
9. They can admit their mistakes and learn from them.
10. They take criticism well.

Qualities that describe people with low emotional intelligence include [11]:

1. They are unable to control their emotions.
2. They cannot maintain good relationships, whether work or personal.
3. They are often emotionally inappropriate for the situation.

**Fig. 6.4** Five elements of emotional intelligence

4. They have trouble coping with sadness.
5. They are emotionally tone deaf and have trouble reading emotions from tone of voice.
6. They have trouble being sympathetic with others.
7. They are completely unmoved by emotional scenes in movies, TV, or books.
8. They trivialize the importance of emotions in general and elevate the importance of "cool, calm logic."

## 6.5 Improving Your Emotional Intelligence

While general intelligence (IQ) is harder to change, emotional intelligence (EQ) is something one can improve on or develop. Our emotional intelligence skills are great contributors to our overall success in life. Emotional intelligence provides "soft skills," which play an important role in our daily lives. Here are four ways for developing EI [12].

*Practice Self-Awareness:* Managing stress is the first step to building emotional intelligence. Self-awareness basically entails understanding yourself and your behavior. Emotional awareness is the ability to identify and name one's own emotions. Spend some time examining your own strengths, triggers, values, and opportunities. There are some EI tests you can take online to help you understand your own EQ.

*Channeling Your Emotions Well:* We must be able to manage our emotions. Emotions are merely indicators that tell us to pay attention to something. Emotions are neither good nor bad, but reactions to emotions could be good or bad.

*Learn to Motivate Yourself:* To be motivated, do not look for inspiration first. Just do *something* that is related to what you want to accomplish. Before you know it, you have made some progress.

*Recognize Emotions in Others to Create Healthier Relationships*: Keep in mind that the whole idea of developing emotional intelligence is to foster healthier relationships in your life. Individuals who are conscious of their emotions and emotions of others have the privilege of enhancing relationships.

There are proven benefits to increasing one's emotional intelligence, including improved mental health, better job performance, and better relationship. By improving and practicing emotional intelligence learning techniques and applying the concepts of EI can help you blaze the trail to successful relationships in all areas of your life.

## 6.6 Applications

Emotional intelligence (EI) has received great attention in many areas generating intense demand for applications in organizations, business, workplace, education, healthcare, and beyond. Emotional intelligence affects your academic performance, your job performance, your professional relationship, your interpersonal communication, your physical health, your mental health, your social relationships, and your social intelligence. We consider some common areas where EI is being applied [13–15].

- *Education:* The concept of emotional intelligence has captured the attention of academia, industry, religious organizations, and the popular media. At a time when education is vital to economic success, research has shown that employers prefer new hires to possess strong EI skills. Education has the goal of preparing students for jobs in particular professions. Emotional intelligence in the academic environment mutually benefits educators and students. It is being applied to different aspects and levels of education. Every profession requires different levels of emotional intelligence. Figure 6.5 shows some exercises that will teach your kids emotional intelligence. These will be of great use throughout their lives.
- *Healthcare*: The role of individuals has been long recognized in the medical profession. Patients undergo emotional distress when faced with news of devastating diagnosis or death. There is the need to develop emotional intelligent healthcare practitioners such as doctors, nurses, therapists, and chaplains who can provide holistic care to the patients. For example, therapists with high EI have the capacity to use relational skills effectively in treatment and should positively influence therapy through their natural abilities. A therapist's memories of personal feelings help to understand a patient experiencing such feelings. Nurses engage in emotional labor as part of their professional work. Emotional labor of nurses may have benefits for both patients and nurses.
- *Business:* Business executives have started to realize that employees can no longer be regarded as machines that come work leaving their feelings and attitudes at home. Every business environment is full of processes and activities with many decision points. Emotions are everywhere in business-related decision environments. Emotions are an important driver in human behavior and decision-making. EI is now regarded as important for businesses to survive, improve, and grow. Business leaders are now embracing the EI concept because of its relationship with self-control, efficiency, productivity, development, decision-making, sales, revenues, quality of service, customer loyalty, greater job satisfaction, employee commitment, and employee morale.
- *Workplace:* Traditionally, emotions have long been a taboo in the workplace. In the workplace, a worker uses social skills, self-awareness, motivation, and self-regulation to acknowledge, comprehend, and regulate emotions. By incorporating emotional intelligence into the workplace, the employees will collaborate better, and the workplace will be a happier environment for everyone. Emotional intelligence also helps in conflict management in the workplace.

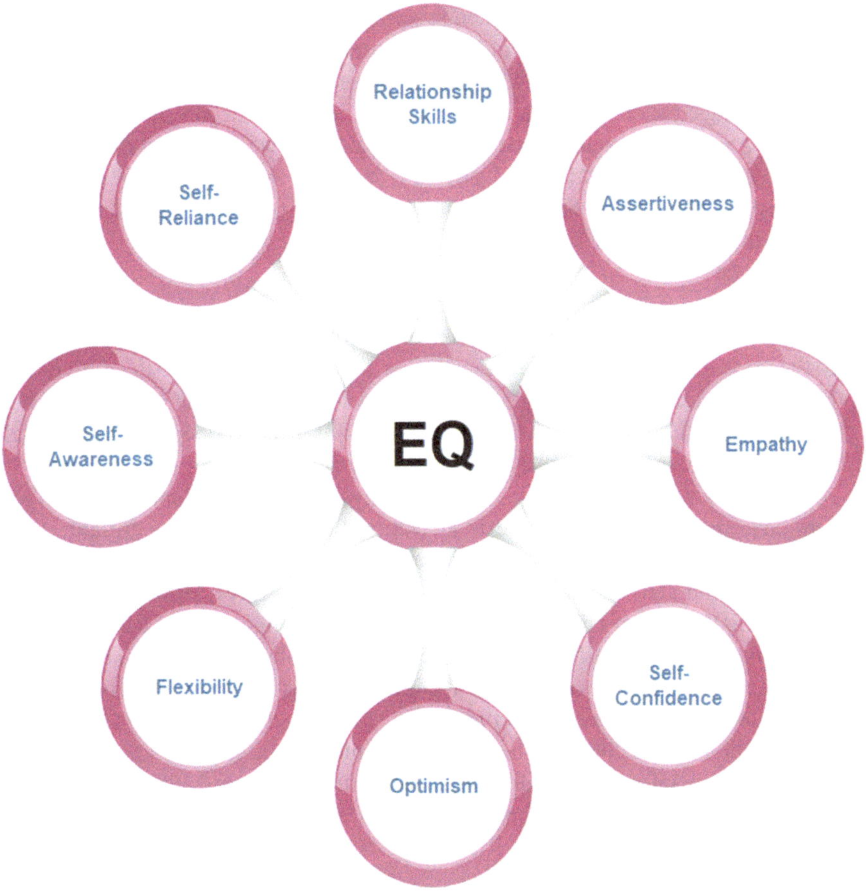


**Fig. 6.5** Some exercises that will teach your children emotional intelligence

Emotional intelligence also enables the employer to respond adaptively to change, nurture trust, and lead teams during tough times. An employer who cannot understand and handle emotions creates problems for their business. Low emotional intelligence at work shows in many ways such as irresponsibility, passive communication, inability to accept constructive criticism, and difficulty in working together as a team.

- *Leadership:* The main purpose of leadership is to influence other humans. David Goleman regarded emotional intelligence as a set of skills and characteristics that drive leadership performance. He indicated that EI accounted for 67% of the abilities deemed necessary for superior performance in leaders. Leaders with high EI would be more aware of their own emotions and from others and could lead companies to better profits and less unnecessary expenses. Having emotional intelligence is crucial for leader success. Emotional intelligence for leadership consists of five components: self-awareness, self-management, empathy, relationship management, and effective communication.

- *Job Performance:* The EI concept has been used in both educational and workplace settings. Studies show a correlation between EI and job performance. EI increases performance and productivity. It can increase the quality of work as well as enhance career success. Under low levels of managerial work demands, a negative relationship between EI and teamwork effectiveness is indicated. Organizations can be successful if they hire people with high emotional intelligence.
- *Job Satisfaction:* Job satisfaction represents a person's evaluation of his or her job and work content. Three factors that affect job satisfaction are: factors related to the job, factors related to the organization, and factors related to the individual. Emotional intelligence is one of the main factors that determine job satisfaction. Including EI in training programs can help employees to cooperate better and motivate more, thereby increasing productivity and profits.
- *Academic Performance:* Possible linkages between EI and academic success factors have been identified. There is a relationship between EI identifiable academic skills, such as grade point average (GPA), attendance, participation in extracurricular activities, and progress towards one's degree. However, evidence supporting the role of emotional intelligence in academic settings is mixed.
- *Relationships:* Relationships are where emotional rubber hits the road. Emotional intelligence plays a crucial role in forming successful human relationships. It facilitates successful navigation through the various situations one goes through in romantic relationships. Higher emotional intelligence is positively correlated with better social relations for adults and better family and intimate relationships. In one study, dating and married couples with higher MSCEIT (the Mayer–Salovey–Caruso Emotional Intelligence Test) scores reported more satisfaction and happiness in their relationships.

Other areas where EI can be applied include management, social work, social networks, entrepreneurship, engineering, children, ethics, communication skills, organizational effectiveness, cooperation/collaboration, conflict resolution, movies, human–machine interactions, artificial intelligence, music, sports, and the list continues.

## 6.7 Assessment of Emotional Intelligence

Lazarus Long once said, "If it can't be expressed in figures, it is not science; it is opinion." Therefore, to study emotional intelligence scientifically requires measuring it. There are several measures of emotional intelligence that have been proposed and used. Such measures generally fall into one of two types: self-report tests and ability tests. Self-report tests are the most common because they are the easiest to administer. Ability tests often require people to demonstrate their abilities. The popular measurement tools include the following [16]:

1. BarOn Emotional Quotient Inventory (EQ-i 2.0)
2. Emotional and Social Competence Inventory (ESCI)
3. Genos Emotional Intelligence Inventory (Genos EI)
4. Group Emotional Competence Inventory (GEC)
5. Mayer–Salovey–Caruso Emotional Intelligence Test (MSCEIT)
6. Schutte Self Report Emotional Intelligence Test (SSEIT)
7. Trait Emotional Intelligence Questionnaire (TEIQue)
8. Work Group Emotional Intelligence Profile (WEIP)
9. Wong's Emotional Intelligence Scale (WEIS)
10. Emotional Competency Inventory (ECI)
11. Emotional Intelligence Inventory (EII)
12. Emotional Intelligence Questionnaire (EIQ)
13. Multifactor Emotional Intelligence Scale (MEIS)

## 6.8 Benefits

General intelligence (IQ) is a measurement of one's ability to process information and come to sound decisions, while emotional intelligence (EQ) is the ability to process emotions. When it comes to success in life, emotional intelligence matters just as much as general intelligence. Emotional intelligence affects your performance at school or work, your physical health, your mental health, your relationships, and your social intelligence. Individuals with a high degree of EI are usually successful in most things they do and are likely to make more money. EI can be beneficial in many areas of life. Other benefits of EI include the following.

- Studies have shown that EI is a key determinant of one's success in life.
- Individuals with high EI have greater mental health, job performance, and leadership skills.
- High EI can make one a better spouse, parent, friend, leader, or citizen.
- EI helps us manage our emotions.
- It increases optimism, confidence, and effectiveness.
- EI fosters good social interactions because it helps you put yourself in other people's shoes.
- EI enhances healthy relationship, whether that relationship is between lovers, friends, family members, church members, or coworkers.
- EI links strongly with concepts of love and spirituality.
- It improves relationships with human beings.
- It improves communication with people.
- It helps you to get respect from others.
- It helps in managing change more confidently.
- It reduces stress levels and increases creativity.

## 6.9 Challenges

Some researchers question the use of the term "intelligence" in "emotional intelligence." Emotional intelligence is vital, but the unbridled enthusiasm has obscured a dark side. Unbridled emotional reactions or lack of emotional intelligence skills by employers and employees can lead to [17]: (1) negative organizational climate/culture, (2) a disengaged workforce, (3) inability or resistance to change, (4) lack of innovation and creativity, (5) Reduced or missed revenue, (5) high cost, (6) decreased productivity, (7) decreased customer satisfaction and customer loyalty, (8) career derailment, (9) workplace violence, (10) increased stress and healthcare costs.

EI has been criticized within the scientific community. Although measuring EI is generally difficult because most of the elements involved are subjective, some effort has been made to measure it [13], such as MSCEIT. But the question is: Are tests of EI reliable? There is no validated psychometric test or scale for emotional intelligence as there is for the general intelligence. Other challenges include the following.

- Some argue that emotional intelligence is trait or skill, not a real intelligence.
- EI is poorly defined and poorly measured.
- EI is defined so broadly and inclusively that it has no intelligible meaning.
- Critics argue that the popularity of EI studies is due to media promotion, rather than rational scientific finding.
- EI models are regarded as inaccurate.
- Like any other intelligence, emotional intelligence can be used for good or evil.

There are some major psychometric and applied issues that must be overcome before EI is regarded as a genuine, scientifically validated concept with real-life practical significance.

## 6.10 Quotes on Emotional Intelligence

The following quotes related to EI will help you reflect and enhance your EI skills [18, 19]:

> Emotions have taught mankind to reason.—Marquis de Vauvenargues

> Emotional self-control – delaying gratification and stifling impulsiveness – underlies accomplishment of every sort.—Daniel Goleman

> We are being judged by a new yardstick; not just how smart we are, or by our training and expertise, but also how well we handle ourselves and each other.—Daniel Goleman

> I don't want to be at the mercy of my emotions. I want to use them, to enjoy them, and to dominate them.—Oscar Wilde

> Feelings can't be ignored, no matter how unjust or ungrateful they seem.—Anne Frank

Be not disturbed at being misunderstood; be disturbed rather at not being understanding.—Chinese Proverb

Running a marathon with a backpack is tough and may hinder you from winning the race. Don't let the baggage from your past – heavy with fear, guilt, and anger – slow you down.—Maddy Malhotra

Anyone can be angry – that is easy. But to be angry with the right person, to the right degree, at the right time, for the right purpose, and in the right way – that is not easy.—Aristotle

The greatest discovery of my generation is that human beings can alter their lives by altering their attitudes of mind.—William James

For every minute you are angry, you lose sixty seconds of happiness.—Anonymous

Angry is a momentary madness, so control your passion or it will control you.—Quintus H. Flaccus

The essential difference between emotion and reason is that emotion leads to action while reason leads to conclusions.—Donald Calne

## 6.11 Conclusion

Emotional intelligence (EI) is a relatively new concept. It can be described as the ability to adaptively perceive, understand, regulate, and harness emotions in the self and others. EI is an emerging concept that has become a part of the fabric of the business community. It is becoming increasingly relevant to organizational development and developing people. EI is now regarded as important for businesses to survive, improve, and grow. Business leaders are now embracing the EI concept because of its relationship with self-control, efficiency, productivity, development, decision-making, sales, revenues, quality of service, customer loyalty, greater job satisfaction, employee commitment, and employee morale. Emotional intelligence is no longer a nice to have for businesses aspiring to be high performing; it is a must-have.

Since Goleman claims that emotional intelligence can be taught, some have suggested that EI should be included in the curriculum (K-12 classrooms and post-secondary) to enhance the success of students in their career [20, 21]. Any form of educational program will be incomplete if it does not cover EI. It will allow students to perform better in academics, leadership, team building, relationship, and decision-making. It is essential to good lawyering, good accounting, good nursing, good engineering, etc.

In spite of its criticism, emotional intelligence has gained wide acceptance among practitioners due to its wide applicability, leading to an outpouring of articles, books, newspapers, magazines, training seminars, and educational curricula. As a result, there are over 10,000 scholarly publications on emotional intelligence.

More information on emotional intelligence can be found in books in [5, 9, 22–44], several other books available on Amazon.com, and the related journals:

- *Cognition & Emotion*
- *Journal of Intelligence*
- *Intelligence*

## References


1. B.K. Patrick, The importance of emotional intelligence, in *Proceedings of the ACM SIGUCCS Fall Conference on User Services*, St. Louis, Missouri, USA (2009)
2. M.N.O. Sadiku, O.D. Olaleye, S.M. Musa, Emotional intelligence assessment. Int. J. Trend Res. Develop. **7**(2), 13–17 (2020)
3. M.N.O. Sadiku, S. Alam, S.M. Musa, Social intelligence: a primer. Int. J. Res. Granthaal. **7**(9), 213–217 (2019)
4. M.N.O. Sadiku, O.D. Olaleye, S.M. Musa, Emotional intelligence: a primer. Int. J. Trend Res. Develop. **7**(1), 109–112 (2020)
5. D. Goleman, *Emotional Intelligence: Why It Can Matter More Than IQ* (Bloomsbury Publishing, London, 1996)
6. Emotional intelligence: the ability model, https://www.universalclass.com/articles/self-help/emotional-intelligence/emotional-intelligence-the-ability-model.htm
7. Chapter 2 Emotional intelligence: an overview, https://shodhganga.inflibnet.ac.in/bitstream/10603/31161/6/06_chapter%202.pdf
8. Karensouthw, Emotional intelligence at work (2011), https://karensouthw.wordpress.com/2011/08/18/emotional-intelligence-at-work/
9. D. Goleman, *Emotional Intelligence: The 10th Anniversary Edition* (Bantam Dell, New York, 2006)
10. Oseni, Reasons you need emotional intelligence to control a business (2019), https://globaltake.com/need-emotional-intelligence-successfully-run-business/
11. C.E. Ackerman, What is emotional intelligence? +18 Ways to improve it, https://positivepsychology.com/emotional-intelligence-eq/
12. M. Manson, 5 skills to help you develop emotional intelligence, https://markmanson.net/emotional-intelligence
13. L.T. Lam, S.L. Kirby, Is emotional intelligence an advantage? An exploration of the impact of emotional and general intelligence on individual performance. J. Soc. Psychol. **142**(1), 133–143 (2002)
14. A. Thiruchelvi, M.V. Supriya, Emotional intelligence and job satisfaction. Asia-Pacific Bus. Rev. **5**, 109–116 (2009)
15. A. Makau-Olwendo, Emotional intelligence and chaplaincy: An analysis of elements of emotional intelligence in chaplains' clinical training and encounters, Doctoral Dissertation, Winebrenner Theological Seminary, Findlay, OH, 2016
16. N.S. Schutte et al., Characteristic emotional intelligence and emotional Well-being. Cognit. Emot. **16**(6), 769–786 (2002)
17. Emotional intelligence introduction and importance, https://static1.squarespace.com/static/548a03c6e4b03faeb7de1f1b/t/57fd3352b3db2b3edb472597/1476211542822/EI+Intro+%26+Importance+2016.pdf
18. C. Moore, Emotional intelligence skills and how to develop them (2019), https://positivepsychology.com/emotional-intelligence-skills/
19. K.O. Sadiku, M.N.O. Sadiku, *Powerful Quotations for Scientists and Engineers* (Triumph Publishing, Bronx, New York, 2017)

20. P. Sherlock, Emotional intelligence in the international education. J. Res. Int. Educ. **1**(2), 139–158 (2002)
21. D.L. Joseph et al., Emotional intelligence training in higher education. New Dir. Teach. Learn. **160**, 51–66 (2019)
22. M. J. Elias, H. Arnold (eds.), *The Educator's Guide o Emotional Intelligence and Academic Achievement: Social-Emotional Learning in the Classroom* (Corwin Press, 2006)
23. J. Ciarrochi, J. R. Forgas, J. D. Mayer (eds.), *Emotional Intelligence in Everyday Life*, 2nd edn. (Taylor & Francis, 2006)
24. T. Badberry, J. Greaves, *Emotional Intelligence 2.0* (San Diego, CA, Talent Smart, 2009)
25. J. Bariso, *EQ Applied: The Real-World Guide to Emotional Intelligence* (Borough Hall, 2018)
26. R.K. Cooper, A. Sawaf, *Executive EQ: Emotional Intelligence in Leadership and Organizations* (Putnam Adult, 1998)
27. D. Goleman, R. Boyatzis, A. McKee, *Primal Leadership: Unleashing the Power of Emotional Intelligence* (Harvard Business Review Press, Boston, MA, 2013)
28. P. Salovey, D.J. Sluyter, *Emotional Development and Emotional Intelligence: Educational Implications* (Basic Books, 1997)
29. R.E. Riggio, S.E. Murphy, F.J. Pirozzolo, *Multiple Intelligences and Leadership* (Psychology Press, New York, 2006)
30. A. D. Fabio (ed.), *Emotional Intelligence – New Perspectives and Applications. Rijeka* (InTech, Croatia, 2016)
31. J. Brauer, *Emotional Intelligence 2.0: Mastery Your Emotions, Develop and Boost Your EQ to Improve the Skill to Business and Relationship* (Independently Published, 2019)
32. B. Goleman, J. Wright, *Emotional Intelligence: For a Better Life, Success at Work, and Happier Relationships. Improve Your Social Skills, Emotional Agility and Discover Why it Can Matter More Than IQ. (EQ 2.0)* (Brandon Goleman, 2019)
33. R. Schulze, R. D. Roberts (eds.), *Emotional Intelligence: An International Handbook* (Hogrefe & Huber Publishers, 2006)
34. J. Ciarrochi, J.H. Beck, *Emotional Intelligence in Everyday Life*, 2nd edn. (Psychology Press, 2013)
35. P. Salovey, M.A. Brackett, J.D. Mayer, *Emotional Intelligence: Key Readings on the Mayer and Salovey Model* (Dude Publishing, 2004)
36. R. J. Sternberg (ed.), *Handbook of Intelligence* (Cambridge University Press, 2000)
37. D. Walton, *Emotional Intelligence: A Practical Guide* (MJF Books, 2016)
38. G. Hasson, *Emotional Intelligence: Managing Emotions to Make a Positive Impact on Your Life and Career* (Wiley, 2014)
39. P.W. Bowen, *Emotional Intelligence: Does It Really Matter? A Guide to Coping with Stressful Experiences* (Vernon Press, 2019)
40. D. Clark, *Emotional Intelligence: The Essential Guide to Improving Your Social Skills, Relationships and Boosting Your EQ (Emotional Intelligence EQ Book 1)* (Amazon.com Services, 2018)
41. T. Bradberry, J. Greaves, *Emotional Intelligence 2.0* (TalentSmart, 2009)
42. P. Salovey, M.A. Brackett, J.D. Mayer, *Emotional Intelligence: Key Readings on the Mayer and Salovey Model* (Due Publishing, Port Chester, NY, 2007)
43. G. Matthews, M. Zeidner, R.D. Roberts, *Emotional Intelligence: Science and Myth* (The MIT Press, Cambridge, MA, 2004)
44. D. Goleman, *Working with Emotional Intelligence* (Bantam Books, New York, 2000)

# Chapter 7
# Interpersonal Intelligence

**Contents**



> I'm more interested in interpersonal relationships - between lovers families, siblings. That's why I write about how we treat each other.
>
> —Terry McMillan

## 7.1 Introduction

The concept of intelligence is interesting and is discussed every day. Intelligence is one's ability to learn from experience and to adapt to environments. It is a property of the mind that includes many related abilities like the capacities to learn, reason, plan, solve problems, think abstractly, and understand ideas. It is a general mental ability to learn and apply knowledge in manipulating the environment. It may also be regarded as the ability that causes a person to function effectively in a certain situation. Intelligence is a vital component of psychology. It is considered as an inborn feature in all human beings [1].

There are different types of intelligence. Interpersonal intelligence is one of nine multiple intelligences of Howard Gardner, a Harvard psychologist [2]. He believed human intelligence was multidimensional. He listed the following forms of intelligence [3]:



M. N. O. Sadiku, S. M. Musa, *A Primer on Multiple Intelligences*,
https://doi.org/10.1007/978-3-030-77584-1_7

1. Musical intelligence: Ability to identify rhythms and sounds.
2. Visual intelligence: Ability to visualize and become aware of the surrounding environment.
3. Linguistic intelligence: Ability to use words effectively.
4. Mathematical/Logical reasoning: Conceptual and abstract thinking.
5. Interpersonal intelligence: Ability of a person to understand and interact with other people effectively.
6. Kinesthetic intelligence: Ability to use the body effectively.
7. Naturalistic intelligence: Unique knowledge about plants and nature.
8. Intrapersonal intelligence: Ability to understand one's innermost feelings.
9. Existential intelligence: The sensitivity and capacity to tackle deep questions about human existence.

Figure 7.1 illustrates the nine multiple intelligences. Evidently, Gardner's list is not exhaustive. There are other types of intelligence such as Social Intelligence, Spiritual Intelligence, Business Intelligence, Emotional Intelligence, Augmented Intelligence, Computational Intelligence, Machine Intelligence, Artificial Intelligence, Abstract Intelligence, Swarm Intelligence, Ambient Intelligence,

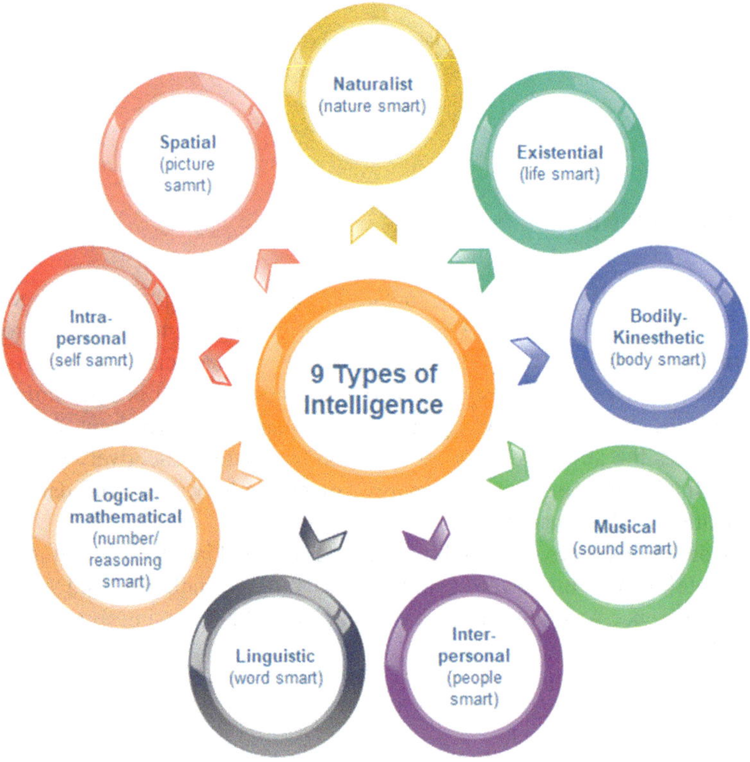


**Fig. 7.1** Illustration of Gardner's theory of multiple intelligences

Digital Intelligence, Military Intelligence, and the list continues. These intelligences are relatively independent. Every intelligence is unique and important in its own backyard.

This chapter provides an introduction to interpersonal intelligence. It begins by discussing what interpersonal intelligence is all about. It presents the characteristics of interpersonal intelligence. It provides some hints on how to develop interpersonal intelligence. It presents some common applications of interpersonal intelligence. It highlights the benefits and challenges of interpersonal intelligence. The last section concludes with comments.

## 7.2 What Is Intepersonal Intelligence?

The term "interpersonal intelligence" was originally developed in 1983 to describe behavior between people in an encounter. Interpersonal intelligence refers to the ability to relate well with others and manage relationships. It is the form of intelligence that governs your ability to relate to others and manage relationships. It may be regarded as the ability to effectively communicate with others, understand and interact effectively with them. It may also be regarded as the ability to recognize distinctions between other people, to react appropriately to their needs, to understand their actions and feelings, and to appreciate such perspectives with empathy. It is a skill we all have to some degree. It is a premium if you are a teacher, politician, pastor, clinicians, therapist, or salesperson. It is useful in school and at work place. People with high interpersonal intelligence are often sensitive, introspective, compassionate, and empathetic [4].

The basic values contained in the main skills of interpersonal intelligence include [5]: (1) the ability to empathize with others, (2) the ability to organize a group of people towards a common goal, (3) the ability to recognize and read the minds of others, and (4) the ability to make friends or establish contacts.

## 7.3 Characteristics of Interpersonal Intelligence

Individuals with interpersonal intelligence have the following interpersonal characteristics [6]:

- Extraverted
- Enjoys social events
- Loves crowds
- Empathetic
- Enjoys teaching others
- Has many friends
- Enjoys team sports

- Counsels others
- Loves meeting new people
- Sensitive to others
- React appropriately to other people's needs
- Enjoy teaching and helping others
- Learn best by working with others
- Have the capacity to understand oneself
- Good verbal communicator
- Are good at discussions and debate
- Good at resolving conflict in groups

These characteristics act as the foundation of interpersonal relationships. Interpersonal intelligence is often mentioned along with intrapersonal intelligence. Although the two are related, they are different. "Intra" means "within," while "inter" means "between." The personal intelligence consists of both intrapersonal and interpersonal intelligence, as shown in Fig. 7.2. Interpersonal intelligence is often called people smart or people quotient (PQ).

People who demonstrate interpersonal intelligence are naturally inclined to become teacher, salesperson, marketing coordinator, communicator, manager, customer service representative, personal services (beautician, nail technician, tattoo artist, etc.), philosopher, researcher, therapist, clergy, social worker, therapists, diplomats, salespeople, receptionist, and negotiators, psychologist, counselor, human resources coordinator, social worker, advisor, attorney, politician, television or radio broadcaster or anchor, actor, nurse, event coordinator, personal trainer, sports coach, recreation therapist, or corporate officer [7]. These are just some of the careers that may use interpersonal intelligence.

According to Gardner, examples of people with high interpersonal intelligence include:

- *Mahatma Gandhi:* He was a lawyer and activist who led India to get independence from British colonial rule. He achieved by employing non-violent civil disobedience. Using interpersonal intelligence, he inspired civil rights and freedom movements across the world.
- *Mother Teresa:* She is best known for her charity work. She helped establish schools for the poor, orphanages, hospices, and leper houses. She founded the

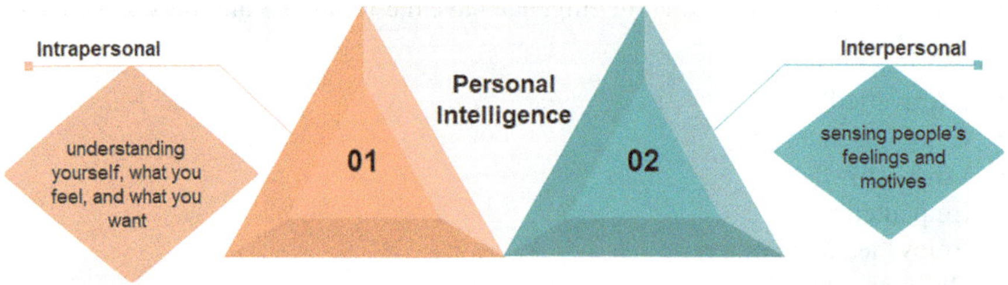


**Fig. 7.2** Personal intelligence (intrapersonal and interpersonal intelligence)

Missionaries of Charity that manages homes for people living with HIV/AIDS, leprosy, TB, and other diseases; orphanages; as well as family and children counseling programs.
- *Oprah Winfrey:* She is a media proprietor, actress, and the host of The Oprah Winfrey Show. She became popular because of her ability to listen, talk, and to relate to others. She became the nation's country's most successful talk show host.
- *Bill Clinton:* He rose from the rank of Governor of Arkansas to become the 42nd President of the United States. His political success has been mainly credited to his personality and ability to relate well with people. Even after his retirement from the presidency, Bill Clinton is involved in public speaking, humanitarian work, and politics.
- *Tony Robbins:* He became self-help coach, motivational speaker, and best-selling author whose seminars have attracted thousands.
- *Phil McGraw:* A psychologist and well-known talk show host, Phil has advised and counseled thousands of people on improving their lives using a tough love approach.

Others include Ronald Reagan, Eleanor Roosevelt, and Bill Gates. These familiar individuals are natural communicators who possess interpersonal skills such as listening, talking, and relating to others [8]. It is good to know that these individuals are not born with interpersonal intelligence.

Lack of interpersonal intelligence will weaken a person's ability to interact with their environment and destroy the relationship with other people. Problems, such as bullying, aggression, withdrawal behavior, and suicidal impulses, are indications of a lower interpersonal intelligence [9].

## 7.4 Developing Interpersonal Intelligence

Now that we understand the importance of interpersonal skills in our lives, let us see how to improve them. There are several ways by which people can enhance, build or develop their interpersonal intelligence. The following ways are suggested [10]:

- Get organized! Use a time management system to make sure you keep in touch regularly with your network of business associates and friends.
- Decide to meet one new person each day (or week).
- Join a volunteer or service-oriented group.
- Start a hobby that involves you having to go to a regular meeting of like-minded people.
- Join the Samaritans.
- Spend 15 min each day practicing active listening with your spouse and children, or a close friend.
- Throw a party and invite people you do not know very well.
- Take a leadership role—at work or in the community.
- Start your own support group.

- Be an appreciative person.
- Participate in workshops or seminars in interpersonal and communication skills.
- Have regular family meetings.
- Become an active listener.
- Communicate clearly.
- Strike up conversation with people in public places.
- Find several pen friends from different cultures and parts of the world.
- Offer to tutor other people on an informal basis in something you have expertise in.
- Spend 15 min a day just people-watching—on a train, bus, shopping center, or restaurant.
- Study the lives of well-known socially competent people and decide what you want to "model" from them.
- Meditate on your connection with those around you—your family, close friends, work colleagues, and community—and eventually, the whole planet!
- Make a list of all the people that you love and make sure you tell them all this week.

The best way to incorporate interpersonal intelligence skills is to work as part of a group and use social networking, social media, and social gaming.

## 7.5 Applications

Interpersonal intelligence refers to the ability of an individual to relate well with people and manage relationships. It is a must-have for professionals such as lawyers and negotiators who will need to look at things from another person's perspective to determine the next action. Interpersonal intelligence is at play in the following applications [11]:

- *Social Interaction:* Individuals with interpersonal intelligence thrive with social interaction. They work well with others, make friends easily, empathize, and understand others. They have natural inner glow and easily attract other people. They know how to keep everyone motivated and happy to work. Consequently, they achieve more with less effort. The higher you go on the career ladder, the more interpersonal intelligence you will need.
- *Influencing Others:* Convincing and persuading others to work together is not easy. With strong interpersonal intelligence, it's possible to influence others to forget about their individual conflicting interests and work for the greater good. We influence others in our speech, life, and example. By being confident and eloquent, team members will look up to you and want to be a part of your end result.
- *Conflict Resolution:* Having strong interpersonal intelligence helps in resolving conflicts and promoting harmony among colleagues. Interpersonal intelligence will help you to objectively analyze situations, come up with a practical solution, and reduce tension among colleagues and hence promote harmony. One needs to

look at the situation from each person's perspective and try to empathize with them. Knowing how to live peacefully with others will greatly benefit your career progress. There are three ways to be an effective conflict resolver. First, create a climate of mutual trust. Second, put the real issue on the table. Third, negotiate win-win solutions [12].
- *The Golden Rule:* This is a maxim of altruism seen in many religions and cultures. It is the principle that encourages one to treat others as one would like to be treated.
- *Education:* Interpersonal intelligence plays a vital role in education. Educating students in the twenty-first century requires a different approach because students are living and learning in a world with ever-changing and increasing technology. Students need to become critical, independent thinkers, problem solvers, good communicators, analyzers of information, and the ability to work with peers. Interpersonal intelligence is needed in order to achieve this [13]. A student displays the characteristics of interpersonal intelligence by getting along with others in the class and working well with others to complete an assignment. Students with interpersonal intelligence showcase their skill in the classroom. They enjoy school group activities such as speech, drama, debate, and social gatherings. Teachers can help such students use their skills through a variety of activities, which will prepare students in future occupations by learning how to solve conflicts, take risks, and empathize with others [14]. Homeschooling parents with large families find ways to help their children in shaping their intellectual development. A parent can help their child with an aptitude for interpersonal skills by matching talents to specific careers.
- *Workplace:* All employees need to be interpersonal intelligence or people smart because everyone in today's workforce deals with people. The people business is no longer the domain of the few. It now includes everyone.
- *Communication:* Interpersonal intelligence involves being able to communicate well with other people. It involves both verbal and nonverbal communication. Interpersonally intelligent individuals can cooperate in order to work as team members. They communicate effectively and empathize easily with others, and may be either leaders or followers. Interpersonal intelligence helps student's oral proficiency.

Some of these applications are illustrated in Fig. 7.3. Other areas of applications of interpersonal intelligence include learning interpersonal skills, languages, translation, and sports.

## 7.6 Benefits

Interpersonal intelligence helps you become friendly and bring unity to a team. It helps to achieve more when working with others and reap maximum benefit. If you are sensitive to the way other people feel, it would be easier for you to make them

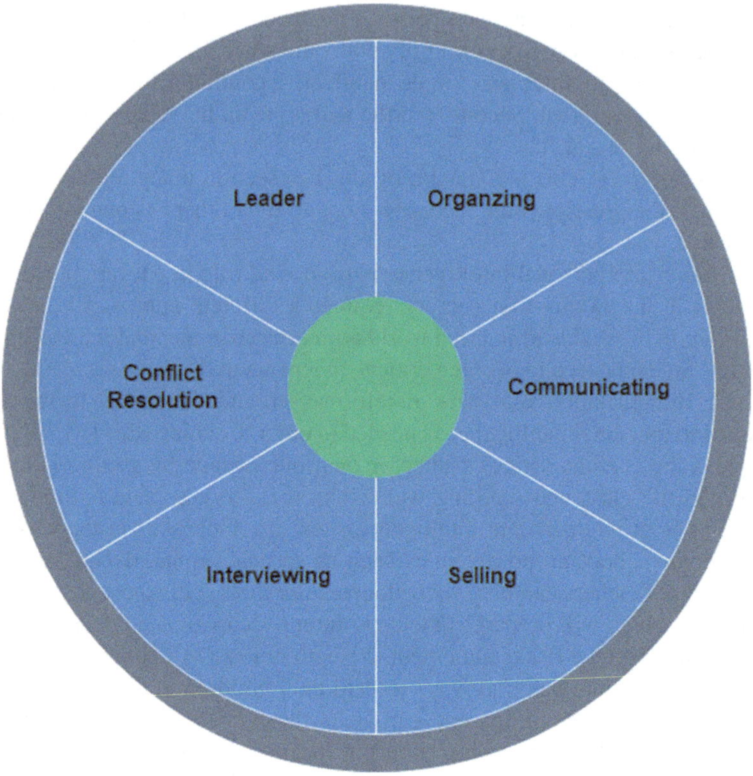


**Fig. 7.3** Typical applications of interpersonal intelligence

do what you desire. Interpersonal intelligence helps you to take advantage of good opportunities in the workplace and it propels you to managerial or supervisory role. Individuals with high interpersonal intelligence are characterized by their sensitivity to other people's moods, temperaments, motivations, and feelings. It is a crucial factor for people in occupations that require interacting with people or leading them. Individuals with high interpersonal intelligence thrive with social interaction. They are strong in communicating with and understanding other people. They have a superb ability to establish rapport with others quickly and easily.

Other benefits of interpersonal intelligence in your career progress include the following [15]:

1. You will be more influential hence people will be happier to work under you.
2. You will have better access to information and new opportunities.
3. You will have an easier time seeking approvals from superiors.
4. You will naturally rise up to fill any leadership vacuum when the situation calls for it.
5. Interviews will be more enjoyable.
6. Networking events will be more useful.

7. People will remember you more, hence when opportunities arise, you will be more likely to be chosen.
8. By empowering and encouraging your colleagues, you will be able to build a stronger team that will produce even better results in your assignments resulting in more rewards for everyone.
9. You will be able to resolve conflicts early before they turn into rivalry that can impact negatively on the group or department's results.
10. You will be able to give feedback without offending, hence it will be easy to make improvements without creating conflicts.

## 7.7 Challenges

There have been a number of questions on human brain and language learning, some of which are yet to be answered. The most common questions concern what makes some people more intelligent than others. Some researchers have questioned interpersonal intelligence in spite of its significance. They have questioned the role of interpersonal intelligence in language learning. Some attribute higher intelligence scores of bi/multilingual speakers to greater mental flexibility and a greater facility in concept formation [16]. English is weak in describing intensities of interpersonal intelligence.

## 7.8 Conclusion

Interpersonal intelligence is essentially how well you interact, work, and deal with other people. It is how well you understand yourself. It is salient in our daily life. It is important regardless of the field, industry, or career. No matter which field you work in, interpersonal intelligence will be beneficial to you in the long run. Interpersonal skills are needed to interact in social situations. Interpersonal intelligence allows us to affect others by understanding their intentions, motivations, personalities, and desires. Without it, we lose the ability to exist socially. For more information about interpersonal intelligence, one should consult the books in [2, 3, 12, 16, 17] and a related journal *Intelligence*.

## References


1. M. Hajebi, S. Taheri, M. Noshadi, The relationship between interpersonal intelligence, reading activity and vocabulary learning among Iranian EFL learners. Int. J. Eng. Lang. Trans. Stud. **6**(1), 48–57 (2018)
2. H.E. Gardner, *Multiple Intelligences: New Horizons in Theory and Practice* (Basic Books, 2007)

3. H.E. Gardner, *Frames of Mind: The Theory of Multiple Intelligence* (Basic Book, New York, 1983)
4. M.N.O. Sadiku, S.M. Musa, A. Ajayi-Majebi, Interpersonal intelligence: an introduction. Int. J. Trend Res. Develop. **7**(4) (2020)
5. S. Sutarman, D. Sunendar, Y. Mulyati, Investigating cooperative learning model based on interpersonal intelligence on language learners skill to write article. Int. J. Instr. **12**(4), 201–218 (2019)
6. Interpersonal intelligence: thriving with social interaction, https://personalitymax.com/multiple-intelligences/interpersonal/#:~:text=People%20with%20Interpersonal%20intelligence%20thrive,and%20often%20have%20many%20friends
7. A. Logsdon, A. Rice, Characteristics of the social interpersonal learning style (2020), https://www.verywellfamily.com/interpersonal-learning-styles-2162780
8. Teaching students identified with interpersonal intelligence, https://www.thoughtco.com/interpersonal-intelligence-8091
9. S.W. Utami, I. Hitipeuw, Emotional, interpersonal and intrapersonal intelligences: the differences between informatics and psychology students. World Trans. Eng. Technol. Educ. **17**(3), 4040–4408 (2019)
10. Inspiring breakthrough, http://www.inspiring-breakthrough.co.uk/learning-styles/interpersonal-learning.htm
11. M. Luenendonk, The role of interpersonal intelligence for career progress (2017), https://www.cleverism.com/interpersonal-intelligence-and-career-progress/#:~:text=It%20helps%20you%20become%20a,maximum%20benefit%20without%20interpersonal%20intelligence
12. M.L. Silberman, M. Silberman, F. Hansburg, *People Smart: Developing Your Interpersonal Intelligence* (Berrett-Koehler Publishers, San Francisco, CA, 2000)
13. D.M. de Nevers, Interpersonal intelligence and problem-based learning, Master's Thesis, Dordt College, April 2014
14. C.A. Dien, D.U. Wustqa, The interpersonal intelligence profile of seventh-grade students in mathematics learning. J. Phys. Conf. Series (2018)
15. R. Mobashshernia, S.M. Aghazadeh, The relationship between the interpersonal intelligence and reading comprehension achievement of Iranian bilingual and multilingual EFL learners. Int. J. Res. Eng. Educ. **3**(2), 13–20 (2018)
16. A. Canny, *Interpersonal Intelligence: Importance of Relating Well to People* (Independently Published, 2019)
17. G. Blokdyk, *Interpersonal Intelligence A Complete Guide* (Emereo Pty Limited, 2020)

# Chapter 8
# Intrapersonal Intelligence

## Contents



> Active use of awareness and learning intrapersonal skills enables us to overcome the current limitations seen in mindfulness and mediation practices. By learning, training and directing intrapersonal skills we can open our inner potential and lead ourselves towards better mental wellness.
>
> —Helena Lass

## 8.1 Introduction

Intelligence refers to an individual's abilities to solve problems and adapt for positive outcomes. The conventional intelligence is based on an aptitude in a very specific set of skills, such as verbal fluency, mathematical proficiency, and logical/analytical ability. Recently, there have been many competing theories of intelligence. More contemporary theories of intelligence indicate that there are many aspects of intelligence that may exist independent of each other.

Intrapersonal ("intra"—"within") intelligence is one of the nine categories of intelligence in the Theory of Multiple Intelligences, developed by Howard Gardner, a Harvard psychologist in 1983 [1]. He shook the educational world by what he called multiple intelligences. He believed human intelligence was multidimensional. He listed the following forms of intelligence [2, 3]:


M. N. O. Sadiku, S. M. Musa, *A Primer on Multiple Intelligences*, https://doi.org/10.1007/978-3-030-77584-1_8

1. *Musical intelligence*: The capacity to think and feel in terms of sound and rhythm.
2. *Visual Intelligence:* The ability to see and represent the world using concepts such as shape, color, and form.
3. *Mathematical/Logical reasoning*: The ability for working with numbers, logical concepts, and abstract analysis.
4. *Interpersonal intelligence*: The ability to understand and interact with other people effectively.
5. *Bodily-Kinesthetic intelligence*: The ability for motor skills, such as dancers, athletes, sculptors, or even surgeons.
6. *Naturalistic intelligence*: The ability to perceive and classify the intricate patterns in the world around them.
7. *Intrapersonal intelligence:* A person's innate ability to understand his or her inner world, a world from which many people are entirely disconnected.
8. *Linguistic intelligence:* The ability to express oneself with such clarity that others can comprehend and relate to them.
9. *Existential intelligence*: The sensitivity and capacity to tackle deep questions about human existence.

Figure 8.1 illustrates nine multiple intelligences. Gardner's principal claim is that, rather than there being one single intelligence, there is a profile of intelligences, each with its own strengths and weaknesses, unique to each individual, which can be nurtured and developed [4]. These intelligences are relatively independent. Every intelligence is unique and important in its own right. People can possess different types and levels of intelligence.

Intelligences can be divided into cool and hot groups. The cool group of intelligences focuses on general and often impersonal information such as verbal meanings or visually patterned information. Hot intelligences operate on information that is more personally relevant and therefore more emotionally charged in nature. The hot group of intelligences includes the social, emotional, and practical intelligences [5]. A person can be successful in more than one intelligence field. In fact, everyone is gifted with all eight intelligences.

This chapter provides an introduction on intrapersonal intelligence. It begins by discussing what intrapersonal intelligence is all about. It discusses the characteristics of intrapersonal intelligence and of intrapersonal intelligent individuals. It provides some actionable tips for developing intrapersonal intelligence. It gives some applications of intrapersonal intelligence. It provides how to assess or measure intrapersonal intelligence. It highlights some benefits and challenges of intrapersonal intelligence. The last section concludes with comments.

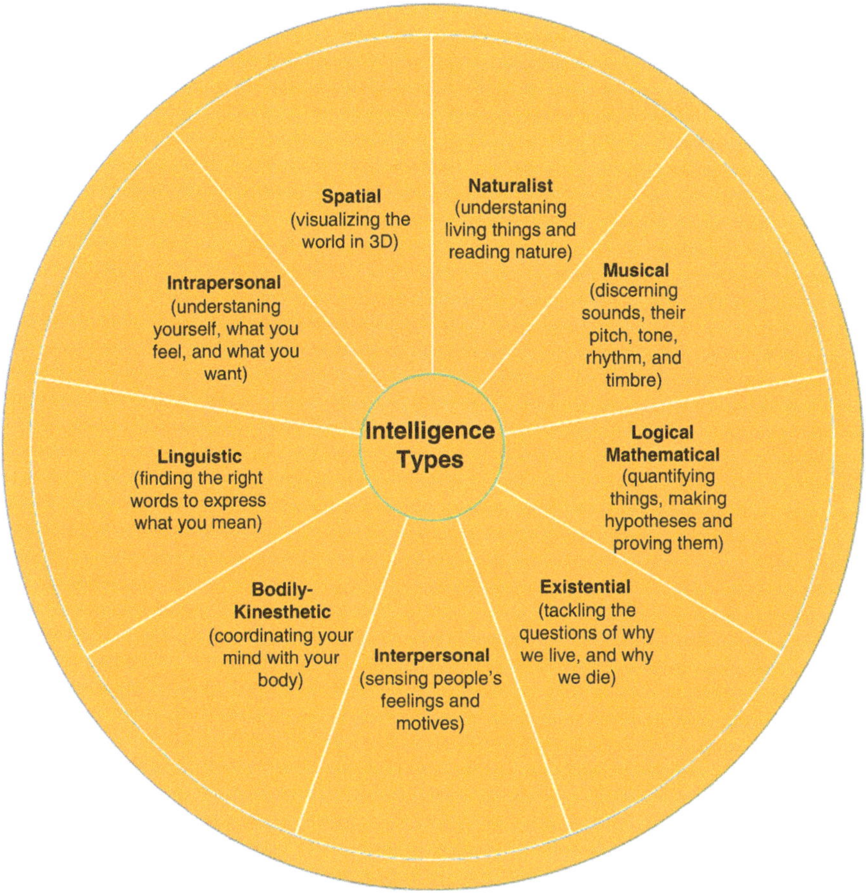


**Fig. 8.1** Illustration of Gardner's nine multiple intelligences

## 8.2 What Is Intrapersonal Intelligence?

Intrapersonal intelligence (self smart) is the capacity to explore one's inner world and feelings. It is the ability to think for yourself without the aid of social processes. It is essentially the ability to see realistically and honestly who we are and what we want.

It is the type of intelligence we exercise when we try to learn more about ourselves. It is the basis for personal productivity, creativity, and learning. It is a form of intelligence we all possess to some degree. Individuals who excel in intrapersonal intelligence are typically self-motivated, introverted, spend lots of time alone, work independently, and achieve a lot.

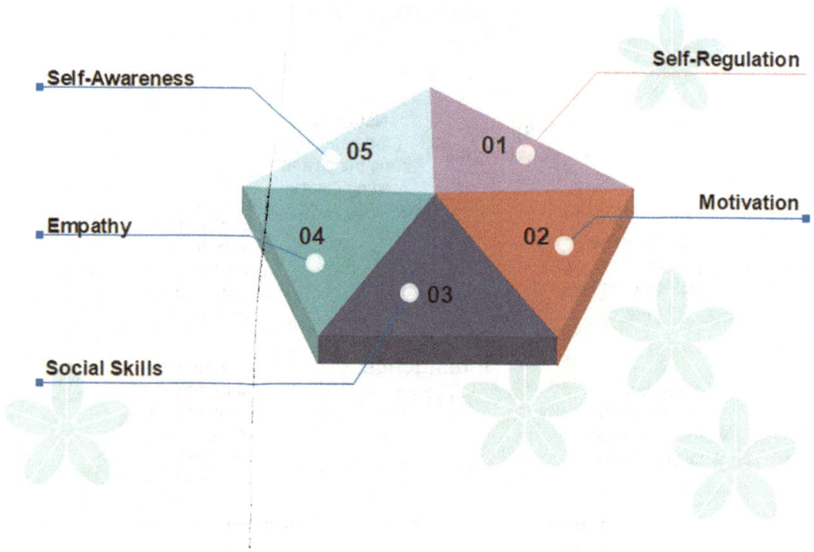


**Fig. 8.2** Some aspects of intrapersonal intelligence

Every effort to discover more about what works and contributes to your success is related to intrapersonal intelligence. As shown in Fig. 8.2, self-awareness, empathy, self-regulation, motivation, and social skills are all aspects of intrapersonal intelligence. Keeping a journal, a diary, and an online blog are all forms of intrapersonal intelligence. People with intrapersonal intelligence tend to shy away from others. They are often quite introspective and are in tune with themselves. Many successful writers, philosophers, professors, and psychologists have great intrapersonal intelligence [6].

Examples of intrapersonal skills include self-esteem, open mindedness, being aware of your own thinking, the ability to learn, being able to understand and manage your own emotions, self-confidence, self-discipline, self-motivation, being able to overcome boredom, being patient, being a self starter, being able to take initiative, working independently, being persistent, having a positive attitude, and being a good manager of time, and the list continues [7].

## 8.3 Characteristics of Intrapersonal Intelligence

Characteristics of interpersonal intelligence include: good at communicating verbally, skilled at nonverbal communication, sees situations from different perspectives. Examples of intrapersonal skills include attributes such as planfulness, self-discipline, mindfulness, self-regulation, delay of gratification, the ability to overcome distractions, and the ability to adjust one's strategy as needed.

The main characteristics of a person with a highly developed intrapersonal intelligence include affective awareness, ethical awareness, self-regulation, and metacognition. Individuals with interpersonal intelligence has the following additional characteristics [8]:

- They can easily apply what they have learned to their own lives.
- They start getting to know other people by getting to know themselves. (Your mantra: To thy own self be true.)
- They enjoy being alone and dealing with something alone.
- They are intuitive and usually introverted.
- Although they are aware of their strengths, they often do not boast about it.
- They set realistic goals for themselves.
- They are independent and have high self-confidence.
- They can control many feelings, especially anger.
- They enjoy analyzing theories and ideas.
- They can easily apply what they have learned to their own lives.
- Young people may be more introverted compared to their peers.
- They like to process the events they have experienced.
- They seek to understand their own mistakes in any negative events they have experienced in order to avoid similar situations in the future.
- They tend to express their feelings and thoughts in creative ways such as through writing or music.
- Development is inevitable for people with high intrapersonal intelligence potential. They can constantly look for new life philosophies and ways to improve themselves.
- They constantly look for new things to learn and new ways to improve themselves.

These characteristics act as the foundation of intrapersonal relationships. Interpersonal intelligence is often mentioned along with intrapersonal intelligence. Although the two are related, they are different and may seem like *polar opposites*. The personal intelligence consists of both intrapersonal and interpersonal intelligence [9]. While intrapersonal ("intra"—"within") is the intelligence of the self, interpersonal ("inter"—"between") is the intelligence of others. Intrapersonal intelligence deals with understanding of self, while interpersonal intelligence deals with understanding of others. Some of the characteristics associated with intrapersonal skills are opposite to those for interpersonal skills. To achieve success one must balance both intrapersonal and interpersonal intelligence.

People who demonstrate interpersonal intelligence are naturally inclined to become entrepreneur, farmer, historian, inventor, librarian, philosopher, psychologist, scientist, theologian, counselor, author, poet, religious leader, police, and researcher.

Examples of individuals with high intrapersonal intelligence include [10, 11]:

- *Aristotle:* This Greek philosopher Aristotle, born 384 BC, is widely credited as the first scholar to study logic. He studied under Plato and had a profound influence on western thought. Socrates, Plato, and Aristotle are celebrated as some of the greatest minds ever. Along with Plato and Socrates, Aristotle was one of the founders of Western philosophy. His dedication to the study of reason required him to examine his own internal motivations, giving him great intrapersonal intelligence.
- *Friedrich Nietzsche:* Aristotle's work made an impact on the nineteenth-century German philosopher Friedrich Nietzsche. He had a widespread influence on social, political, and cultural movements throughout the nineteenth and twentieth centuries. Nietzsche wrote about the forms of spiritual metamorphoses necessary to lead a meaningful life.
- *Albert Einstein:* This is one of history's greatest scientists. Einstein liked to spend time thinking during long walks. His deep thinking sharpened his intrapersonal intelligence. Like Einstein, people with high intrapersonal intelligence are self-motivated, introverted, spend lots of time alone, and work independently.
- *Mahatma Gandhi*: This is an exemplar of intrapersonal intelligence. He was a revolutionary leader and creator of the now universal concept of civil disobedience movement. He believed in the philosophy of non-violence.
- *Mother Teresa:* Her intrapersonal skills led her to become a caring and selfless teacher. Mother Teresa served all her life for the needy, poor, and unwanted with empathy and compassion.
- *Anne Frank:* She demonstrated one of the characteristics of intrapersonal intelligence. She enjoyed writing in journals during tragic circumstances. Before her 1945 death at age 15 during the Holocaust, she spent much of World War II hidden in an attic with her family. While in hiding in German camp, Anne wrote a journal that remains one of the world's most widely known books. Figure 8.3 shows Anne Frank as a good example of intrapersonal intelligence [11].
- *Eleanor Roosevelt:* She was an outspoken politician and humanitarian (a trait is related to intrapersonal intelligence). She was regarded as controversial for her leadership role in the civil rights movement. She was the first US president's wife who used mass media to address issues on women's rights and human rights.

Others include Confucius, Sigmund Freud, Martin Luther King, Jr. and Steve Jobs. By virtue of their intrapersonal intelligence, these famous people paved way for their success and helped a lot of people. They changed the world through exercising their intrapersonal intelligence. Although some people seem to have a natural knack for intrapersonal intelligence, this skill can also be taught and developed.

**Fig. 8.3** Anne Frank as a good example of intrapersonal intelligence [9]

## 8.4 Developing Intrapersonal Intelligence

Everyone has personal strengths and weaknesses. It is expedient to develop areas where one is weak. Intrapersonal intelligence can be developed like other skills.

You can develop intrapersonal intelligence, or knowledge of yourself, by following these actionable tips [12].

- *Keep a Journal:* Journaling can promote intrapersonal intelligence. Your journal provides a great space to keep track of important goals, emotions, and your overall daily experience.
- *Know Yourself*: Spending time and effort in knowing oneself is one of the best things you can do to improve your intrapersonal skills. Know your drives, motivations, frustrations, inspirations, and urges. Write your own autobiography.
- *Practice Meditation:* Meditation can help you focus your attention more effectively. Do it at home, at any time you are comfortable with.
- *Time for Self-reflection:* Taking time for self-reflection can help increase intrapersonal intelligence. It involves thinking over your actions, choices, and the things important in your life.
- *Read Books:* Study scriptures and philosophy. Study the biographies of great individuals you admire.
- *Setting Goals: Set realistic goals according to your needs, and develop motivation and a strong will to achieve them.*
- *Develop social hobbies:* Some individuals may have difficulty adapting to social settings. More involvement among social communities can help manage and eliminate social anxiety.
- *Take Chances:* Risk-taking that ends in success can boost self-confidence.

- *Have Reminders:* Having reminders of what you need to do can help keep you on track despite distractions.
- *Never Give Up*: Never give up on your dreams or goals.

Everybody and anybody can develop high intrapersonal intelligence. Improving your intrapersonal intelligence can help you become a better partner, parent, friend, and coworker.

## 8.5 Applications

Interpersonal intelligence is at play in the following applications [13, 14]:

- *Self-Care:* Taking care of ourselves is more important than taking care of others. If our health suffers, we cannot help others. Self-care may include eating well, getting enough rest, exercising daily, learning to say no, and be careful how we spend our time. You do not have to be everything to everyone.
- *Giftedness:* This is often equated with special ability to reach an expert level much earlier than normatively expected. It entails how individuals use abilities for constructive purposes. True giftedness lies in contribution, what difference or impact individuals make in the world. Strong intrapersonal intelligence is particularly important for those who aim to make a moral contribution to the world.
- *Purpose:* Purpose may be conceived of as an aspect of intrapersonal intelligence. It represents giftedness as an extraordinary and sometimes precocious achievement in the moral use of intrapersonal intelligence. It is an internal moral compass that integrates engagement in activities that affect others and self-awareness of one's reasons.
- *School*: A structured academic environment can help one grow faster in intrapersonal intelligence than working entirely on your own. Teachers can help students enhance and strengthen their intrapersonal intelligence by having assign students independent projects and having them write reflections on the topics covered in class. All of these activities will help students increase their intrapersonal intelligence. Children with high intrapersonal intelligence often score higher in other types of intelligence. A parent should help their kids see the array of occupations that are related to each of their intelligence areas. Figure 8.4 illustrates a solitary learner displaying intrapersonal intelligence [15].
- *Workplace*: The workplace is where you spend most of your time. Change is taking place everywhere including in today's workplace. Knowledge is growing exponentially as technology continually transforms the way we live and work. The routine jobs of yesterday are being replaced by technology or shipped offshore. They have replaced jobs that require knowledge management, abstract reasoning, and personal services. If you are well connected with your self, you will find it hard to connect well with others at work.
- *Mindfulness:* Intrapersonal intelligence is a form of intelligence we all possess, but we must develop it for our benefit. One way to achieve this is through the

**Fig. 8.4** A solitary learner displaying intrapersonal intelligence [13]

practice of mindfulness. Mindfulness is the process of viewing and responding to situations rather than reacting to them. Becoming more mindful will help in making better decisions, identifying what we enjoy, and improve our relationships with others.

- *Communication:* Intrapersonal intelligence can assist and enable effective communication skill. Talking, listening, and writing are different aspects of communication. Communication is a two-way street that involves listening as well as being heard.

Other areas of applications of intrapersonal intelligence include journaling, appreciation, self-effectiveness, self-regulation, self-learning, language learning, animal care, career development, procrastination, and sensitivity.

## 8.6 Assessment of Intrapersonal Intelligence

Discovering your intrapersonal skills will definitely give you an edge both at home and at office, because once you are aware of your latent strengths and weakness you tend to act accordingly. You can use the following intelligence questionnaire to measure your intrapersonal intelligence [16]. Respond on each item using the following five-point scale:

1 = strongly disagree.
2 = disagree.
3 = neither disagree nor agree.
4 = agree.
5 = strongly agree

1. You are well aware of both your strengths and weaknesses.
2. You are very independent and self-directed. You do not like being told what to do.
3. You like learning more about yourself and your inner psychology.

4. You trust your instincts and you are good at following them.
5. You are a unique, original person—and you like being that way.
6. If you are feeling bad, you are usually able to discover the root of your negative emotions.
7. You spend a lot of time thinking about life and reflecting on your place in the world.
8. You prefer to undertake projects alone. You like flying solo.
9. You have a very philosophical outlook on life—and it is always evolving.
10. You are ambitious and confident in your own abilities.

The test above is simple. A more advanced measure of intrapersonal intelligence is the Multiple Intelligences Development Assessment Scales (MIDAS).

## 8.7 Benefits

Intrapersonal intelligence (understanding of self) is the ability to remain productive and creative when you face problems, difficulties, and failures. It gives you a greater compassion and understanding of your own. It enables you to give yourself a chance to really listen to "the voice within." You can embrace your feelings, fears, and motivations without hesitation. Intrapersonal intelligence will help combat self-doubt and increase your professional confidence. Individuals with high intrapersonal intelligence are often intuitive and introverted. They learn independently and through reflection. These people are highly motivated and are often determined to achieve their goal [17].

Intrapersonal intelligence helps students recognize their strengths and weaknesses, which is essential for setting goals. Athletes, coaches, and other professionals use intrapersonal communication to prep-up themselves before major performances. Today, intrapersonal skills are just as important as interpersonal skills.

## 8.8 Challenges

The task of defining intelligence has difficulties and is controversial among educators and psychologists. Robert J. Sternberg has warned that there is no scientific evidence to back up Gardner's theory of multiple intelligences and what he calls intelligences are really aptitudes or abilities (https://www.picuki.com/tag/Intrapersonalintelligence).

It is hard to know exactly how many types of intelligence there are. Although it is expedient to educating young people to find a purpose in life, many youths never reach this intrapersonal integration at all. They figure out what they want or need, but not what they can contribute [14]. Also, we design a standard intelligence test or an intrapersonal intelligence test that can quantify or measure these types of intelligence. There are limitations of an "intelligence" measurement which focused only on some types of intelligence.

## 8.9 Conclusion

Intrapersonal intelligence ("access to one's own feeling life") is basically how we understand ourselves and our emotions, moods, desires, and inspiration. This type of intelligence has to do with knowing one's inner thoughts, emotions, and feelings. It helps us to be conscious of who we are, what we want to do, how to react, what not to do and more. Intrapersonal intelligence shares a lot in common with emotional intelligence. Those who possess intrapersonal intelligence prefer reflection to action. With their determination and self-discipline, they excel at tasks that require a high level of focus. For more information about intrapersonal intelligence, one should consult the books in [1, 2, 18] and a related journal *Intelligence*.

## References

1. H.E. Gardner, *Multiple Intelligences: New Horizons in Theory and Practice* (Basic Books, 2006)
2. H.E. Gardner, *Frames of Mind: The Theory of Multiple Intelligence* (Basic Book, New York, 1983)
3. D. Gutierrez, K. Perri, A. Quackenbush, Exploring the multiple intelligences of community college students enrolled in online courses. J. Coll. Teach. Learn. **3**(11), 85–90 (2006)
4. J.G. Mowat, The development of intrapersonal intelligence in pupils experiencing social, emotional and behavioural difficulties. Educ. Psychol. Pract. **27**(3), 227–253 (2011)
5. J.D. Mayer, Personal intelligence. Imagin. Cogn. Pers. **27**(3), 209–232 (2007–2008)
6. M.N.O. Sadiku, P.O. Adebo, A. Ajayi-Majebi, S.M. Musa, Intrapersonal intelligence: a primer. Int. J. Trend Res. Develop. **7**(5) (2020)
7. Tai, What are intrapersonal skills and why are they important? (2019), https://matterapp.com/blog/what-are-intrapersonal-skills-and-why-are-they-important/#:~:text=Intrapersonal%20skills%20are%20all%20about,easily%20navigate%20your%20interpersonal%20relationships
8. What is intrapersonal intelligence? (2019), https://www.mentalup.co/blog/what-is-intrapersonal-intelligence-examples-and-activities#:~:text=Intrapersonal%20intelligence%20is%20the%20capacity,planning%20and%20managing%20their%20life
9. Howard Gardner's theory of multiple intelligences, https://www.thebettertoystore.com/multiple-intelligences
10. The best examples of intrapersonal intelligence, https://www.thoughtco.com/intrapersonal-intelligence-profile-8092
11. 8 Famous and unforgettable people with intrapersonal intelligence, https://historyplex.com/famous-people-with-intrapersonal-intelligence
12. M.E. Gleason, Intrapersonal intelligence strategies in the developmental writing classroom. J. Virgin. Commun. Coll. **16**(1), 95–105 (2011)
13. You know what intrapersonal intelligence is and why it's so helpful? (2018), https://exploringyourmind.com/intrapersonal-intelligence/

14. S. Moran, Purpose: giftedness in intrapersonal intelligence. High Abil. Stud. **20**(2), 143–159 (2009)
15. Intrapersonal intelligence (AKA self smart), http://blog.kidzmet.com/core-beliefs/multiple-intelligences/intrapersonal-intelligence/
16. How does your intrapersonal intelligence rate? https://www.blogthings.com/howdoesyourintrapersonalintelligenceratequiz/
17. Intrapersonal Intelligence: adept at looking inward and understanding self, https://personality-max.com/multiple-intelligences/intrapersonal/
18. A. Tiwari et al., *iNTELLYJELLY- Interpersonal and Intrapersonal Intelligence: Intelligent Reading Is Fun* (Prerna Publication, Delhi, 2019)

# Chapter 9
# Cultural Intelligence

## Contents



> Cultural Intelligence: An outsider's seemingly natural ability to interpret someone's unfamiliar and ambiguous gestures the way that person's compatriots would.
>
> —Harvard Business Review

## 9.1 Introduction

Culture is everything about our way of living. It is made of factors including language, society, economy, customs, history, and religion. Culture generally shapes people's behaviors, backgrounds, values, and beliefs. It represents how groups organize their knowledge, tangible and intangible values, and behavior, distinguishing one group from another. It is developed by previous generations and transferred to next generations to be further developed and transferred. It subconsciously guides and affects our behavior and thoughts, and often influences our sense of belonging, motivation, and effectiveness at work. Cultures are formed of geography, faith, gender, generation, and organization.

As shown in Fig. 9.1, it is helpful if you can think about culture in the following five ways: (1) Culture is learned, (2) Culture is shared, (3) Culture is dynamic, (4) Culture is systemic, and (5) Culture is symbolic. People who are detached from



M. N. O. Sadiku, S. M. Musa, *A Primer on Multiple Intelligences*,
https://doi.org/10.1007/978-3-030-77584-1_9

their own culture can more easily adopt the mores of an unfamiliar host. Adapting to a new culture involves overcoming obstacles and setbacks. By adopting people's culture, you come to understand what it is like to be them. You show by your actions that you have entered their world. They, in turn, become more welcoming, trusting, and open.

People fail in intercultural situations in all sorts of ways, such as the following [1]:

- Being unaware of key features and biases of our own culture.
- Feeling threatened or uneasy when interacting with people who are culturally different.
- Being unable to understand or explain the behavior of others who are culturally different.
- Trying to apply knowledge about one culture to a different culture.
- Being unable to adjust to living and working in another culture.
- Being unable to develop long-term relationships with people from other cultures.

We live in a world with rapidly increasing connectivity across all businesses and individuals. Big problems can no longer be solved by just one individual, one culture, one company, or even one continent, operating alone. Our world is shrinking and requiring leaders from different cultures to work together. The world is yearning for innovation: new ways, new ideas, new processes, new technologies, and new ventures. Innovation needs people who proactively seek to encourage difference. Today, many businesses can be described as global. Globalization is becoming

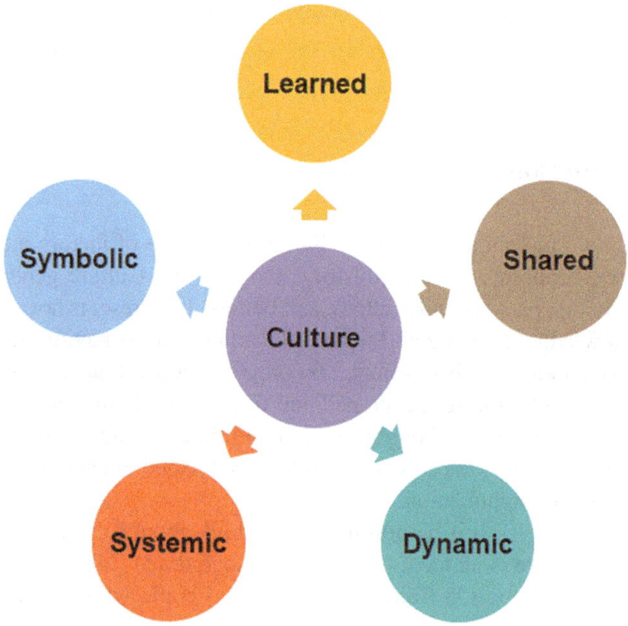


**Fig. 9.1** Elements of culture

instrumental for business success. It has compelled individuals and businesses to come together and interact with each other from different cultures. In busines, it is culture that ultimately enables a competitive advantage to sustain and grow over time [2, 3].

Intelligence refers to the ability to respond flexibly to new situations, to learn, and to innovate. Its definition is culture bound. There are individual differences in human intelligence, attributable to cultural background. Cultural differences also define one's intelligence; what is considered "smart" in one community might be considered "foolish" in another. Cultural intelligence is the capability to relate and work effectively across cultures. It is about understanding ourselves and understanding others. It is a crucial skill in today's world. It is now more important than ever in our daily lives due to increasing globalization and growing diversity.

This chapter introduces the reader to cultural intelligence. It begins by discussing what cultural intelligence means. It explains the four elements of cultural intelligence and how to develop it. It presents some applications of cultural intelligence. It highlights some benefits and challenges of cultural intelligence. It addresses the related issue of intercultural intelligence. It provides some quotes on cultural intelligence. The last section concludes with comments.

## 9.2 What Is Cultural Intelligence?

Cultural intelligence (also known as cultural quotient or CQ) refers an individual's capability to function effectively in a culturally diverse setting. It is the ability to make sense of unfamiliar contexts and then blend in. Cultural intelligence is a form of interpersonal intelligence. It is distinct but related to emotional intelligence. Both emotional intelligence and cultural intelligence are subsets of social intelligence. Cultural intelligence is somewhat related to emotional intelligence; it picks up where emotional intelligence leaves off. An element they share in common is to think before acting.

Cultural intelligence may also be regarded as the ability to recognize and adapt to cultural differences. It can give you the confidence to operate successfully in a wide range of cultural settings. CQ is becoming a significant aspect of global business, offering a host of benefits to companies that embrace it. It is a must-have skill to work across borders.

The term "cultural intelligence" is relatively recent. It was introduced in 2003 by Christopher Earley in his book [4]. Since then, CQ has attracted worldwide research and study. Cultural intelligence builds on insights from intelligence research suggesting that intelligence is multifaceted. It is an ability that varies with individuals and explains differences in intercultural effectiveness. It can be learned and developed by almost anyone.

Cultural intelligence consists of three components: knowledge, mindfulness, and behavioral ability, as depicted in Fig. 9.2. These three components combine to produce the ability to interact effectively across cultures. Cultural intelligence resides in the body and the heart, as well as the head. Thus, three key components of CQ may also be identified as Head, Body, and Heart.

- *Head:* This is the knowledge and understanding that you need good CQ. This comes, in part, from observation and research.
- *Body:* This means translating cultural information into visible actions.
- *Heart:* To have high CQ, you need to be self-assured, not afraid to make honest mistakes, and confident enough to keep improving.

Adapting to a new culture involves overcoming obstacles and setbacks. People with high CQ use all three of these components to monitor their actions. They can interpret what is going on in any cultural setting and adjust their behavior accordingly.

Low CQ can turn out to be a disadvantage. Individuals with high emotional intelligence can pick up on the emotions, wants, and needs of others. People with high cultural intelligence are attuned to the values, beliefs, attitudes, and body language of others from different cultures. Cultural intelligence includes the abilities to understand, manage, and deal with the emotions of others.

## 9.3 Elements of Cultural Intelligence

We conceptualize cultural intelligence into four critical elements: Drive, Knowledge, Strategy, and Action. Your cultural intelligence is made up of these four different capabilities. According to author David Livermore, culturally intelligent people exhibit the four capabilities [5–7]:

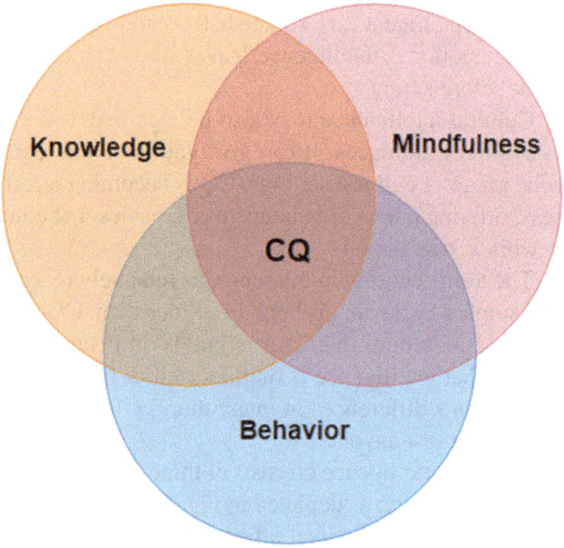


**Fig. 9.2** Three components of cultural intelligence

1. CQ-Drive: This is the motivation to learn about new cultures. It is a person's interest and confidence in functioning effectively in culturally diverse settings. It includes intrinsic interest, extrinsic interest, and self-efficacy. Drive is what motivates you to learn about and respond to a different culture. Confidence is needed in CQ, because unfamiliar settings can be daunting and challenging. To enhance your CQ-Drive, make effort to interact with people in different communities, learn a foreign language, learn the history of a culture, and be actively involved in projects that connect you with people from different cultures.
2. *CQ-Knowledge:* In order to do business with people from different cultures, we need to know what makes them tick and what is important to them. CQ-knowledge is a person's knowledge about how cultures are similar and how cultures are different. It includes business, interpersonal knowledge, and socio-linguistics. Cultural knowledge is about knowing how that culture shapes people's behaviors, background, values, and beliefs.
3. *CQ-Strategy:* This refers to leaders' level of awareness and ability to strategize across cultures. It takes having a plan to respond to cultural differences. This is how a person makes sense of culturally diverse experiences. It includes awareness, planning, and checking.
4. *CQ-Action:* This is a leader's ability to act appropriately in a range of different cultural settings. It involves knowing when to adapt to another culture and when not to. It is behaving in culturally sensitive ways. This is a person's capability to adapt verbal and nonverbal behavior to make. It shows how you react when things do not go according to your plan.

These elements are illustrated in Fig. 9.3. By focusing on these four elements, CQ offers a comprehensive framework that describes the domain of intercultural capabilities. Leaders with high CQ have strength in four distinct areas. Sometimes a 20-item cultural intelligence scale (CQS) is used to measure all four factors of cultural intelligence. This is one way of measuring CQ.

## 9.4 Developing Cultural Intelligence

Different people achieve different levels of cultural intelligence based on experience, education, and personality. Cultural intelligence can be developed in many ways such as formal intercultural education, living in other cultures, visiting other cultures, and having personal relationships with people from different cultures and backgrounds [8]. Anyone who is reasonably alert, motivated, and poised can attain an acceptable CQ. Developing your CQ helps you to be attuned to the values, beliefs, and attitudes of people from different cultures. Attitude is crucial in developing and improving cultural intelligence. Cultural intelligence can be developed in psychologically healthy people by following these steps [9].

Step 1: The individual examines his CQ strengths and weaknesses in order to establish a starting point for subsequent development efforts.

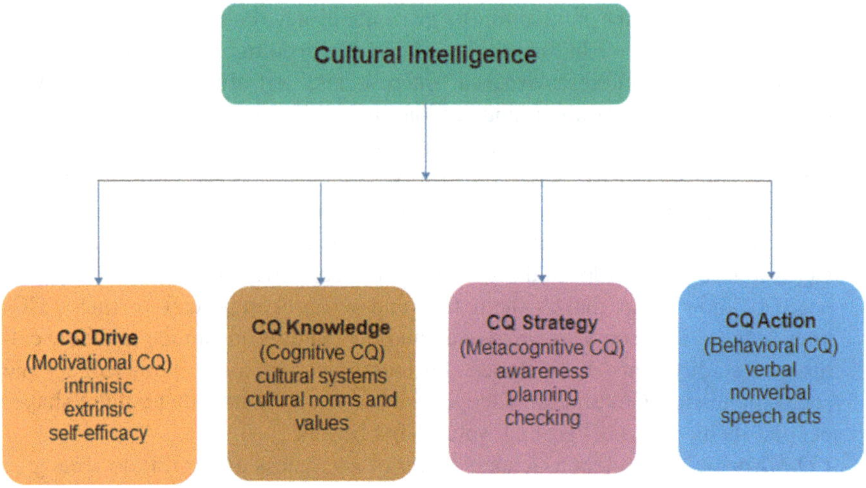


**Fig. 9.3** The four elements of cultural intelligence

Step 2: The person selects training that focuses on his weaknesses.

Step 3: The individual organizes her personal resources to support the approach he has chosen.

Step 4: The person enters the cultural setting he needs to master.

Step 5: The individual reevaluates his newly developed skills and how effective they have been in the new setting.

## 9.5 Applications

Cultural intelligence is a person's capability to function effectively in situations characterized by cultural differences. The concept of cultural intelligence, whether cultivated or innate, is used in business, government, workplace, and leadership. It is applied in the following areas [5, 7].

- *Business*: CQ has been gaining acceptance within the business community.
- Figure 9.4 shows that CQ is one of the six dimensions of intelligence contributing to organizational success. The community has expanded the application of cultural intelligence from the individual to the organizational construct. CQ teaches that allowing knowledge and appreciation of the difference to guide responses will lead to better business practice. The idea of cultural intelligence has been expanded into a capability that yields a commercial advantage by turning cultural trends into profits. Businesses that break down cultural barriers and promote mutual acceptance perform better than those that do not. A business can harness the benefits of diversity and inclusion. Culturally intelligent leaders cre-

**Fig. 9.4** Cultural intelligence is one of the six dimensions of intelligence contributing to organizational success

ate an environment where diversity and culture thrive. Cultural intelligence is important in tailoring marketing campaigns to diverse audiences. With the ongoing globalization and increasing diversity, people belonging to different cultural backgrounds are working together to ensure business success. American companies have created hiring strategies that support the new diversity requirements in business. For example, banking and finance is a global industry. People working in this industry should be able to understand and communicate effectively across cultures.

- *Workplace*: Today's workplaces are multicultural. Cultural intelligence in the workplace is all about TRUST: Tolerance, Respect, Unity, Solidarity, and Teamwork [10]. We work with colleagues who have cultural backgrounds, assumptions, and attitudes that are very different from our own. Managers are expected to navigate through the thicket of habits and assumptions that define their employees' differences. Foreign cultures are everywhere—in other countries, certainly, but also in corporations, college campuses, vocations, and regions. Interacting with individuals within them demands perceptiveness and adaptability. For diversity and culture to thrive in a company, everyone in the workplace must hold each other accountable for their choices and behaviors. Everyone is encouraged to perform his best and to hold each other to the highest standards. Figure 9.5 illustrates workplace diversity and inclusion [11].
- *Leadership:* Leadership involves complex interactions of leaders and followers who may have diversity of needs, skills, and abilities. Cultural intelligence is a topic of urgency for organizational leaders. Today, leadership is a multicultural challenge. Leaders are individuals who inspire, motivate, unite people, and create visions for the future. Cultural intelligence is the must-have skill for any

modern leader because it offers the advantage that leaders need in a world where people increasingly work across borders, both physical and culturally. Businesses now benefit from global mobility and need leaders with high cultural intelligence to be successful. Globalization has put those who are in leadership positions into a situation where they must interact with multicultural customers, suppliers, and employees on a daily basis. Leaders in multicultural environments must create a culture of mutual interdependence. For them to lead effectively, leaders need cultural intelligence (CQ), intelligence quotient (IQ), and emotional intelligence (EQ). Leadership without cultural intelligence in a rapidly globalizing world leads to hardship, frustration and confusion, poor job performance, decreased revenues, poor working relationships, and lost business opportunities.

- *Government*: The application and integration of cultural intelligence into the local government improve the effectiveness of governments to respond to and serve a diverse population. Cultural intelligence is central to military intelligence and ensuring successful military operations. Governmental negotiation and

**Fig. 9.5** Culturally intelligent workers [11]

diplomacy can be made much more effective if knowledge of the peoples is understood.

- *Education:* Our campuses are increasingly global and diverse. Teachers today face the reality of the No Child Left Behind Act. This presents challenges for teaching effectiveness. Teachers need strategies and tips for creating an inclusive classroom environment and teaching a culturally diverse population of students. A culturally inclusive classroom is one where students and teachers appreciate and capitalize on diversity so as to enrich the overall learning experience [12].
- *Artificial Intelligence:* These are machines that can meet or beat human performance at a specific cognitive task. In order to deploy AI agents widely in our community (on our highways, in our nursing homes, schools, businesses, governments, etc.), we will need machines that are wise and smart. AI agents cannot afford to ignore cultural rule because doing so may annoy the group and lead them to fear or reject the machine [13].
- *Military:* Culture has become a significant aspect of military operations. Military personnel are often deployed to foreign countries and are expected to quickly forge collaborative relationships in a multinational environment. It is apparent that cultural intelligence is critically important to the operational effectiveness of the military. In recent years, researchers have investigated the usefulness of CQ to better prepare military personnel for overseas deployment [14]. The United States military's ability to fight the Global War on Terror (GWOT) also requires that it be capable of operating in diverse environments. To be successful in their mission, military personnel must be able to bridge cultural gaps in distant unfamiliar lands. Military intelligence is needed at three levels: strategic, operational, and tactical. Cultural intelligence serves the dual purpose of supporting peace support operations as well as in peacekeeping [15].
- *Tourism:* Tourism is associated with tolerance, multiculturality, cultural exchange, and creation of good friendship. There is need for cross-cultural education to prepare tourism students in a modern, multicultural century. Cultural intelligence is a promising advancement in tourism [16].

Other areas of applications of cultural intelligence include management, conflict resolution, career success, cross-cultural skills, job satisfaction, diplomacy, hotel industry, design industry, and international students' adaptation.

## 9.6 Assessment of Cultural Intelligence

Cultural intelligence or CQ is the ability to cross boundaries and thrive in multiple cultures. It can be measured on a scale, similar to that used for measuring an individual's IQ. For example, you can use the following CQ test to measure your cultural intelligence [17]. Respond on each item using the following five-point scale:

1 = strongly disagree.

2 = disagree.

3 = neither disagree nor agree.
4 = agree.
5 = strongly agree

1. Before I interact with people from a new culture, I ask myself what I hope to achieve.
2. If I encounter something unexpected while working in a new culture, I use this experience to figure out new ways to approach other cultures in the future.
3. I plan how I am going to relate to people from a different culture before I meet them.
4. When I come into a new cultural situation, I can immediately sense whether something is going well or something is wrong.
5. It is easy to change my body language (for example, eye contact or posture) to suit people from a different culture.
6. I can alter my expression when a cultural encounter requires it.
7. I modify my speech style (for example, accent or tone) to suit people from a different culture.
8. I easily change the way I act when a cross-cultural encounter seems to require it.
9. I have confidence that I can deal well with people from a different culture.
10. I am certain that I can befriend people whose cultural backgrounds are different from mine.
11. I can adapt to the lifestyle of a different culture with the relative ease.
12. I am confident that I can deal with a cultural situation that is unfamiliar.

The test above is simple. A more advanced measure of cultural intelligence is the Cultural Intelligence Scale (CQS) developed by Earley and Ang in 2003. The scale seeks to measure an individual's ability to understand, act, and manage effectively in culturally diverse settings. It is a 20-item scale used to measure all four factors of cultural intelligence: (1) Motivational, (2) Cognitive, (3) Metacognitive, and (4) Behavioral [14]. Besides the popular CQS, we also have the Intercultural Development Inventory (IDI) and the Short Form measure of Cultural Intelligence (SFCQ).

## 9.7 Benefits

Cultural intelligence is the capability to cross boundaries and prosper in multiple cultures. It is about taking time to understand and adapt to different cultural norms. It is beneficial to develop cultural intelligence as it implies greater access to more resources, knowledge, and talent. It helps you to work effectively and get along easily with anyone from a different culture and background. It can give you the confidence to operate successfully in a wide range of settings. Cultural intelligence is a consistent predictor of performance in multicultural settings. Professionals with high CQ are more successful on international assignments. You need CQ to run a business today because CQ will help you make more money.

In today's globalized world, individuals and organizations must be culturally intelligent. Cultural intelligence is crucial for those of us in foreign countries and those who regularly travel to other countries to work. Businesses with culturally intelligent and diverse employees will be more likely to accomplish their goals in today's globalized economy. Other benefits of cultural intelligence include [18–20]:

- CQ allows you to be attuned to the values, beliefs, and attitudes of people from different cultures and work effectively with them.
- Culturally intelligent individual gains trust by assimilating into a local culture.
- Companies that embraces diversity, reinforces inclusion, and promotes cultural sensitivity will be more appealing than traditional businesses that are perceived as inflexible and old-fashioned.
- It adds a competitive edge to a business, as it boosts the day-to-day operation of a business and improves communication, cooperation, teamwork, and job performance.
- It helps people to develop an in-depth understanding of working styles in other cultures.
- It encourages practices that support diversity, equity, inclusion, and belonging.
- It helps in communicating effectively in different cultural settings.
- It establishes a common ground.
- It maximizes teamwork in groups where people act and think differently.
- It provides tools for improving sales success, people performance, and brand image.
- It encourages people to be the best they can be.
- It offers insights to better understand international colleagues and clients.
- It enables better collaboration in both remote and face-to-face teams.
- It increases creativity and innovation.

## 9.8 Challenges

Although some people are genuinely willing to interact and learn from other cultures, ultimately there is a lack of substantial engagement. Missed opportunities to learn from one another can be seen in universities around the world. This is especially true for international students, who are potential global leaders. Bridging cultural differences can make or break a business.

Google is one of several large foreign companies that have failed in China because they made some basic cultural mistakes. Many international charities, businesses, and government programs are ineffective and lose a lot of money because they are unable to develop their practice of cultural intelligence. The performance of business leaders with low cultural intelligence impacts their profitability. The same could be said of politicians or military leaders who think that CQ is of little importance [21].

Although culture can enhance long-term success, many executives do not make it a priority. The development of CQ between employees can be complicated, time-consuming, and costly. The organizational risks associated with a lack of cultural intelligence skills include miscommunications, critical business errors, lack of team cohesion and trust, demotivation, high staff turnover, loss of clients, and an impact to the bottom-line [22]. Some critics doubt if assessment of cultural intelligence is credible. Their objections include: (a) cultures are said to be entirely relative in their values; (b) cultural studies are said to be a form of postmodernism; and (c) attempts to categorize cultures are said to be crude stereotypes lacking subjectivity [23].

## 9.9 Intercultural Intelligence

Intercultural intelligence refers to the capability to function effectively in culturally diverse settings and consists of different dimensions: metacognitive, cognitive, motivational, and behavioral. Intercultural intelligence differs from cultural intelligence in that it is based on the belief in interculturalism while CQ is based from the belief in multiculturalism. Intercultural Intelligence is used in the business world to bridge the cultural divide and avoid misunderstandings that affect business. As businesses become global, so do the interactions among their multicultural employees that are managed across cultural lines. Intercultural intelligence has been integrated into schools and colleges [24].

## 9.10 Quotes on Cultural Intelligence

> Cultural Intelligence: An outsider's seemingly natural ability to interpret someone's unfamiliar and ambiguous gestures the way that person's compatriots would—Harvard Business Review

> Culture is more often a source of conflict than of synergy. Cultural differences are a nuisance at best and often a disaster.—Geert Hofstede.

> What people say, what people do, and what they say they do are entirely different things.—Margaret Mead

> A people without the knowledge of their past history, origin and culture is like a tree without roots.—Marcus Garvey

> People of different religions and cultures live side-by-side in almost every part of the world, and most of us have overlapping identities which unite us in very different groups.—Kofi Annan

> No culture has ever developed all human potentialities; it has always selected certain capacities, mental and emotional and moral, and shifted others. Each culture is a system of values which may well complement the values in another—Ruth Benedict

> Most of my important lessons about life have come from recognizing how others from a different culture view things—Edgar H. Schein

## 9.11 Conclusion

Cultural Intelligence (CQ) is an emerging concept. It is the ability to recognize and adapt to cultural differences. It consists of a set of skills to relate and work effectively in culturally diverse situations. It is measured on a scale similar to that used to measure an individual's intelligence quotient. The greater your cultural intelligence, the more likely you will successfully adjust to the cultures you encounter. Teams with low CQ underperform compared to those with high levels when it comes to innovation, engagement, performance, and profitability. Individuals with high levels of CQ are better able to accomplish goals in an effective way regardless of the cultural context.

Businesses and organizations in the twenty-first century need culturally intelligent managers and leaders to survive. Globalization has increased the cross-cultural interaction and created the need for knowledge and competence about different culture. This has led to an extraordinary growth in cultural intelligence research. Cultural intelligence is the number one predictor of success in today's borderless world. For more information on cultural intelligence, one should consult the books in [4, 6, 23–39] and the journals that are exclusively devoted to it:

- *Intelligence.*
- *Studies in Intelligence.*
- *American Intelligence Journal.*
- *International Journal of Intercultural Relations.*

## References

1. D.C. Thomas, Measuring cultural intelligence: implications and opportunities. Rutg. Bus. Rev. **2**(2), 205–211 (2017)
2. M.N.O. Sadiku, O.D. Olaleye, A. Ajayi-Majebi, S.M. Musa, Cultural intelligence: a primer. Int. J. Trend Res. Develop. **7**(4) (2020)
3. What is cultural intelligence? https://commonpurpose.org/knowledge-hub-archive/all-articles/what-is-cultural-intelligence/
4. P.C. Earley, S. Ang, *Cultural Intelligence: Individual Interactions Across Cultures* (Stanford University Press, Palo Alto, 2003)
5. Cultural intelligence, Wikipedia, the free encyclopedia, https://en.wikipedia.org/wiki/Cultural_intelligence

6. D. Livermore, *The Cultural Intelligence Difference: Master the One Skill You Can't Do Without in Today's Global Economy* (AMACOM, 2011)
7. L. Brancu, V. Munteanua, I. Goleta, Understanding cultural intelligence factors among business students in Romania. Procedia. Soc. Behav. Sci. **221**, 336–341 (2016)
8. M.T.G. Carranza, C.P. Egri, Managerial cultural intelligence and small business in Canada. Manag. Rev. **21**(3), 353–371 (2010)
9. P.C. Earley, E. Mosakowski, *Cultural Intelligence* (Harvard Business Review, 2004)
10. N. Cherradi, What are the advantages of cultural intelligence in business? https://www.moroccoworldnews.com/2017/03/211497/advantages-cultural-intelligence-business/
11. How culturally intelligent professionals thrive in a 21st-century workplace, https://go.chatwork.com/blog/en/2016/10/20161020why-culturally-intelligent-professionals-thrive-in-a-21st-century-workplace.html
12. G. Hadfield, Safe artificial intelligence requires cultural intelligence (2018), https://techcrunch.com/2018/09/11/safe-artificial-intelligence-requires-cultural-intelligence/
13. D.F. O'Keefe, D.Y. Bourgeois, K.D. Davis, Measuring cultural intelligence: psychometric evaluation of the cultural intelligence scale with a military sample in the Canadian armed forces, http://resmilitaris.net/ressources/10254/78/res_militaris_article_o_keefe_et_alii_measuring_cultural_intelligence_in_the_canadian_armed_forces.pdf
14. A. Ellis, The influence of cultural intelligence (CQ) on faculty leadership, Doctoral Dissertation, Florida Atlantic University, Boca Raton, FL, 2017
15. H. Yalçınkaya, Y. Özer, Another lesson learned in Afghanistan: the concept of cultural intelligence. Int. Peacekeep. **24**(3), 434–460 (2017)
16. M.K. Bobanovic, J. Grzinic, Teaching tourism students with cultural intelligence. UTMS J. Econ. **10**(1), 85–95 (2019)
17. CQ TEST, http://nbsmlt.nb.ca/userfiles/file/CQ%20Test%20Apr%202014.pdf
18. S. Bryant, 12 Reasons why more cultural intelligence is needed in the finance industry (2019), https://www.tmaworld.com/our-thinking/12-reasons-why-more-cultural-intelligence-needed-finance-industry
19. M. PazFollow, Cultural intelligence and its importance in the workplace (2015), https://www.linkedin.com/pulse/cultural-intelligence-its-importance-workplace-margot-paz
20. 7 Benefits of improving your cultural intelligence (2016), https://www.amanet.org/articles/7-benefits-of-improving-your-cultural-intelligence/
21. Driving global leadership with cultural intelligence, https://davidcliveprice.com/global-leadership-agility/
22. Intercultural intelligence, Wikipedia, the free encyclopedia, https://en.wikipedia.org/wiki/Intercultural_intelligence
23. C. Hampden-Turner, F. Trompenaars, *Cultural Intelligence: Is Such a Capacity Credible?* (Group & Organization Management, 2006)
24. M. Blankenburgh, *Inter-Cultural Intelligence: From Surviving to Thriving in the Global Space* (Book Baby, 2013)
25. D. Livermore, *Leading with Cultural Intelligence: The Real Secret to Success*, 2nd edn. (AMACOM, New York, 2015)
26. B. Peterson, *Cultural Intelligence: A Guide to Working with People from Other Cultures*, 2nd edn. (Across Cultures, 2019)
27. S. Ang, L. V. Dyne (eds.), *Handbook of Cultural Intelligence: Theory, Measurement, and Applications* (Routledge, 2015)
28. D.C. Thomas, K. Inkson, *Cultural Intelligence: People Skills for Global Business* (ReadHowYouWant, 2009)
29. R.D. Bucher, *Building Cultural Intelligence (CQ): Nine Megaskills* (Pearson, 2009)
30. D. Livermore, *Cultural Intelligence (Youth, Family, and Culture)* (Baker Academic, 2009)
31. D.C. Thomas, K. Inkson, *Cultural Intelligence: Surviving and Thriving in the Global Village* (Berrett-Koehler Publishers, 2017)
32. E. Plum, *Cultural Intelligence: The Art of Leading Cultural Complexity* (Libri Publishing, 2009)

33. Y. Liao, D.C. Thomas, *Cultural Intelligence in the World of Work: Past, Present, Future* (Springer, 2020)
34. M. Moua, *Leading with Cultural Intelligence* (Saylor Foundation, 2012)
35. D. Livermore, S. Ang, *Leading with Cultural Intelligence: The Real Secret to Success* (AMACOM, 2015)
36. P.C. Earley, S. Ang, *Cultural Intelligence: Individual Interactions Across Cultures* (Stanford Business Books, 2003)
37. P.C. Earley, S. Ang, J.S. Tan, *CQ: Developing Cultural Intelligence at Work* (Stanford Business Books, Stanford, CA, 2006)
38. B. Peterson, *Cultural Intelligence: A Guide to Working with People from Other Cultures* (Intercultural Press, 2004)
39. M.D. Kim, *Preaching with Cultural Intelligence* (Baker Academic, 2017)

# Part II
# Machine Intelligence

# Chapter 10
# Machine Intelligence

**Contents**



> We need machines, but more than that we need humans who know how to use those machines for the greater good.
>
> —Abhijit Naskar

## 10.1 Introduction

The traditional machine was a physical system that lacked intelligence. It was just a simple machine that was intended to substitute human's physical labor [1]. The idea of creative thinking, reasoning machines has been appealing and fascinating to human imagination. Several human activities (such as engaging in conversation or driving a car) requires intelligence.

We have associated intelligence with capabilities such as solving difficult problems, reasoning consistently and reliably, and processing information quickly. All our activities as human beings can be described by five components: data, prediction, judgment, action, and outcomes. Machine intelligence (MI) is essentially a prediction technology. It lowers the cost of goods and services that rely on prediction. This is important since prediction is an input to a host of activities including transportation, agriculture, healthcare, energy manufacturing, and retail [2].

The chapter provides an introduction to machine intelligence. It begins by discussing what machine intelligence is all about. It provides some applications of


M. N. O. Sadiku, S. M. Musa, *A Primer on Multiple Intelligences*,
https://doi.org/10.1007/978-3-030-77584-1_10

machine intelligence. It addresses some benefits and challenges of machine intelligence. The last section concludes with comments.

## 10.2 What Is Machine Intelligence?

The idea that an artificial brain could compete with that of human brain has been around for many years. Researchers worldwide have attempted to put human intelligence into machines by the ways of algorithms. They have focused on the goal to engineer machines that can act in ways that are similar to the way humans act.

Machine intelligence refers to the intelligence exhibited by computers. It results from the development of artificial intelligence (AI) and is often presented as artificial intelligence. Machine intelligence is basically a prediction technology. Since prediction can have a profound effect on our life, we often make efforts to predict our future from the past and the present. Adaptability is a vital component of any intelligent machine; it has already been introduced in several automatic devices. The goal of machine intelligence is giving machines the ability to learn from the changing environment and making predictions [3].

There is no exact and universally accepted definition of intelligence in humans or machines. An intelligent machine that can perform tasks would be helpful in almost any human endeavor. Larry Fogel identified three key elements of intelligence [4]: (1) ability to predict, (2) ability to adapt, and (3) ability to take appropriate action.

We can also consider intelligence as consisting of mainly communication and learning [5]:

1. *Ability to communicate:* It is natural to expect an intelligent machine to be able to communicate in human language. Human knowledge and thinking are always mediated using natural languages. The intelligent machine can be a computer that is programmable. It must be able to communicate in an interactive manner.
2. *Ability to learn*: An intelligent machine must be able to learn. Learning enables the machine to adapt and modify itself based on experience and external environment. There are several types of behavior that can be regarded as learning. Machine learning can be related to control theory and artificial intelligence. Today machine learning requires large amounts of data.

As shown in Fig. 10.1, machine intelligence has two components—control intelligence and interface intelligence. Control intelligence performs control in response to unanticipated events in uncertain environments. Interface intelligence indicates the degree of intelligence of the human–computer interaction (HCI).

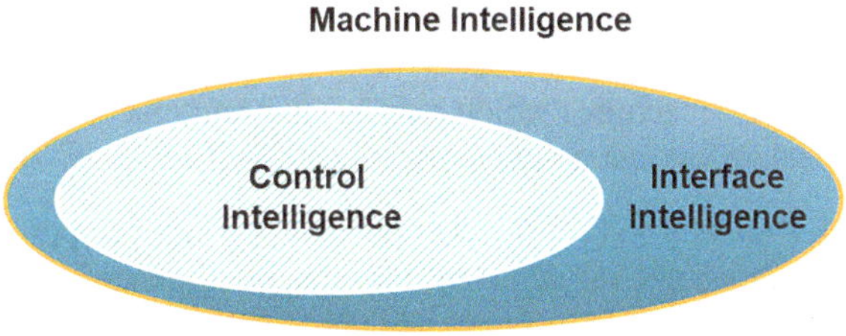


**Fig. 10.1** Two components of machine intelligence

## 10.3 Relationship Between MI, AI, and ML

The concept of machine intelligence forms the intersection of machine learning and artificial intelligence. Machine intelligence, artificial intelligence, and machine learning are related but distinct from each other. When a machine is programmed to learn, it looks for patterns and trends within the given data and uses those findings to make conclusions.

Artificial intelligence (AI) will take in information to come up with new solutions using all aspects of human intelligence. AI might match or even exceed human intelligence and capabilities on tasks such as complex decision-making, reasoning and learning, sophisticated analytics and pattern recognition, speech recognition, computer vision, natural language processing, and language translation. It will surpass the human mind in several ways than imaginable [6].

Machine learning (ML) is a set of computational tools that use data to create models that make predictions about future data. It requires massive amounts of data from which patterns can be recognized and predictions can be made. ML addresses a challenge that is faced by traditional AI, which is essentially the inexact nature of real-world rules and patterns.

Machine intelligence (MI) is created when machines are programmed to perform some aspects of human intelligence such as learning, reasoning, strategizing, problem-solving, and prioritization. This implies that machine intelligence has a battery of different machine learning techniques. MI relies on ML and combines imitation (i.e., artificial intelligence), self-learning, extrospection, and introspection.

## 10.4 Applications

Common applications of machine intelligence include dynamic programming, nonlinear dynamic, gradient search techniques, fuzzy logic, Markov processes, chaos, opinion-guided reaction, conditioned responses, neural networks, and stimulus reinforcement.

There are now machines which show adaption and learning; machines which can classify speech and hand-writing; machines which can perform goal-directed motor activity; machines which play games; and machines which can communicate [7].

Here are just a few applications of MI.

- *Automation and Robots*: Automation can be approached from the point of view of machine intelligence. Machine intelligence is enabling applications we already expect like automated assistants (Siri), adorable robots (Jibo), and identifying people in images. Another example is the success of Microsoft's chatbot Little Bing, to which millions of Chinese confide their deepest secrets. Robotics is undergoing explosive growth due to improvements in computational power, new materials, and better sensors. For example, car manufacturers use automation in form of welding and assembly robots. Figure 10.2 shows the transition from robotics and AI to machine intelligence.
- *Games*: Machine intelligence is used in games. The Checkers program CHINOOK was the first official man-machine world champion in any game in

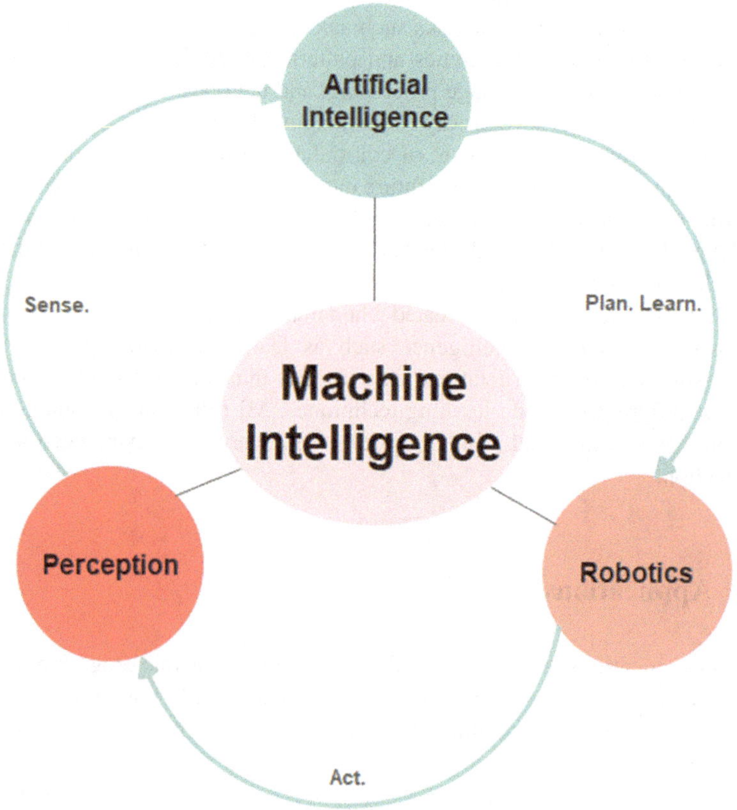


**Fig. 10.2** From robotics and AI to machine intelligence

1994. Hex is another challenging zero-sum board game that involves two players [8]. Watson was developed by IBM as a question answering computer system capable of answering questions posed in natural language.

- *Space Exploration*: Through telecommunication technology, humans can control space vehicles at remote places. The concept of teleoperation is important in space program. NASA has identified five directions for developing intelligent machines [9]: (1) automated operations in space, (2) precision pointing and control, (3) efficient data acquisition, (4) real-time data management, and (5) low-cost data distribution.
- *Artificial General Intelligence*: The domains of AI include Artificial Narrow Intelligence (ANI), Artificial General Intelligence (AGI), and Artificial Super Intelligence (ASI). The concept of general intelligence as involving cognitive flexibility aligns well with contemporary definitions of intelligence in computer science. There is the possibility of developing autonomous AGI. AGI has not come close to being achieved. Work on AGI projects around the world includes OpenCog, an open-source AGI framework; the Non-Axiomatic Reasoning System (NARS); and the Learning Intelligent Distribution Agent (LIDA).
- *Cognitive computing:* This is the sensory branch of machine intelligence. Cognitive computing brings human capabilities to computers so they can think, behave, and act like humans. Using sensors and algorithms, computers can "see," "hear," and "feel." Cognitive computing is the way laymen experience artificial intelligence. Cognitive computing makes computers think, behave, and act like humans by adding artificial sensory capabilities to computers and adding a brain to computers [10].

Other application areas of machine intelligence include healthcare, pattern recognition, automatic target recognition (ATR), and modern photogrammetry.

## 10.5 Machine Intelligence Quotient

By nature, scientists want to measure every thing or concept. To be able to make informed decisions on MI, we need to understand how to measure intelligent machines for their machine intelligence. Measuring "machine intelligence" is hard. There is no absolute measure, only relative. Some researchers have taken a number of well-known informal definitions of human intelligence and then mathematically formalized these definitions to produce a general measure of intelligence for arbitrary machines. Some have introduced Machine Intelligence Quotient (MIQ) as a new index to represent machine intelligence [11]. Multiple perspectives, machine intelligence measurement (MIM), and fuzzy set theories are used to determine the commonalities and differences among the current diverse machine intelligence quotient (MIQ) theories.

## 10.6 Benefits

Machines have been designed to have the ability to learn from experience, detect patterns, organize information, identify relationships, make predictions, and detect anomalies. Machine intelligence is the technology that is very human, enough to inspire shock and awe. Machines that perform tasks once only possible for humans is already a reality. Such machines will enrich and improve human lives. Companies who will win the machine intelligence race will be more powerful than they are right now.

Today's machines are learning to think, read, and write. They are also picking the ability to see, touch, taste, and smell. Nowadays, computers that can speak in human language get to know us better, intimately. They are capable of performing narrowly defined tasks with great precision.

## 10.7 Challenges

Although the benefits of machine intelligence are clear, building these intelligence machines presents significant challenges in terms of risks and costs. In today's world, human intelligence is still far superior to machine except in the ability to process, compute, store, and distribute data at fast rates. For example, machine intelligence never independently produced an original idea or discovered any new principle. An intelligent machine has never come up with a solution to a social problem or a crime [12]. The possibility of human-equivalent general intelligence is still a science fiction and may not happen in our lifetime. In spite of the progress in machine intelligence, artificial general intelligence is still a major challenge.

Sometimes the best way to use machine intelligence is to pair it with human intelligence. The success of MI is determined by human–machine interaction (HCI). For example, the HCI factor determines the success of modern learning systems, from Google Car to the digital assistant. Some perceive that although intelligent machines can serve as useful tools, they can also be used for dubious purposes. The same can be said of any human invention such as computers, vehicles, guns, and robots [13]. There are ingrained prejudices against machine intelligence.

Since the beginning of the Industrial Revolution, workers have been uprooted progressively by new machines. However, the new machines cannot replace the engineers who invented, designed, and produced them [14]. Thinking machines appear as a threat for humanity. There is a shortage of talent in machine intelligence.

## 10.8 Conclusion

Google is at the forefront of innovation in machine intelligence. MI at Google raises deep scientific and engineering challenges, allowing us to contribute to the broader academic research community through technical talks and publications.

Learning and applying knowledge is a crucial to human intelligence. Collective intelligence that builds on the human intelligence is different from Turing machine intelligence. The intelligent machines should have the ability to learn, adapt, and predict [15]. Machine prediction is cheaper and better substitute for human prediction, just as machines outperform humans in arithmetic. As MI lowers the cost of prediction, we will start using it as an input for things for which we never previously did. The need for skilled professionals in machine intelligence is continuously on the rise. More information about machine intelligence can be obtained from books in [7, 13, 16–30] and the journals devoted to it: *Nature Machine Intelligence* and *Intelligence.*

## References

1. S. Wang, H. Di, Machine intelligence and intelligent machines, in *Proceedings of 2011 International Conference on Fluid Power and Mechatronics* (2011), pp. 487–491
2. A. Agrawal, J. Gans, A. Goldfarb, The simple economics of machine intelligence (2016), https://hbr.org/2016/11/the-simple-economics-of-machine-intelligence
3. M.N.O. Sadiku, D.S. Adesina, S.M. Musa, Machine intelligence. Int. J. Adv. Res. Comp. Commun. Eng. **9**(2), 51–53 (2020)
4. Z. Michalewicz, M. Michalewicz, Machine intelligence, adaptive business intelligence, and natural intelligence, in *IEEE Computational Intelligence Magazine* (2008), pp. 54–63
5. T. Mikolov, A. Joulin, M. Baroni, A roadmap towards machine intelligence, https://arxiv.org/pdf/1511.08130.pdf
6. What exactly is machine intelligence? (2019), https://www.forbes.com/sites/quora/2019/11/15/what-exactly-is-machine-intelligence/#5c1211d0187c
7. N. L. Collins, D. Michie (eds.), *Machine Intelligence* (Oliver & Boyd, Edinburgh, 1967)
8. S.K. Chalup, D. Mellor, F. Rosamond, The machine intelligence Hex project. Comput. Sci. Educ. **15**(4), 245–273 (2005)
9. Machine Intelligence and Robotics: Report of the NASA Study Group (1980), http://www.rr.cs.cmu.edu/NASA%20Sagan%20Report.pdf
10. 4 Types of machine intelligence you should know (2018), https://www.informationweek.com/big-data/ai-machine-learning/4-types-of-machine-intelligence-you-should-know/a/d-id/1331480?
11. H.J. Park, B.K. Kim, G.Y. Lim, Measuring machine intelligence for human-machine cooperative systems using intelligence task graph, in *Proceedings of the 1999 International Conference on Intelligent Robots and Systems* (1999), pp. 689–694
12. M. Lee, *10 Questions for Machine Intelligence* (The Futurist, 2014), p. 64
13. D. Beyer, *The Future of Machine Intelligence* (O'Reilly Media, Sebastopol, CA, 2016)
14. K.A. Toth, The workless society: how machine intelligence will bring. The Futurist **24**(3), 32–37 (1990)
15. J. Ma, S. Chen, Y. Xu, Fuzzy logic from the viewpoint of machine intelligence. Fuzzy Sets Syst. **157**, 628–634 (2006)

16. B. Kosko, *Neural Networks and Fuzzy Systems: A Dynamic Systems Approach to Machine Intelligence* (Prentice Hall, Englewood Cliffs, 1991)
17. R. Sharma, *Machine Intelligence in Design Automation* (Independently Published, 2018)
18. D.B. Fogel, *Evolutionary Computation: Toward a New Philosophy of Machine Intelligence*, 3rd edn. (Wiley-IEEE Press, 2005)
19. W.H. Land Jr., J.D. Schaffer, *The Art and Science of Machine Intelligence: With An Innovative Application for Alzheimer's Detection from Speech* (Springer, 2020)
20. G. Carlsson, F.X. Campion, *Machine Intelligence for Healthcare* (CreateSpace Independent Publishing Platform, 2017)
21. K. Hosanagar, *A Human's Guide to Machine Intelligence: How Algorithms Are Shaping Our Lives and How We Can Stay in Control* (Viking, 2011)
22. H. S. Nwana, N. Azarmi (eds.), *Software Agents and Soft Computing Towards Enhancing Machine Intelligence: Concepts and Applications* (Springer, 1997)
23. P. S. P. Wang (ed.), *Pattern Recognition, Machine Intelligence and Biometrics* (Springer, 2011)
24. H. He, *Self-Adaptive Systems for Machine Intelligence* (Wiley, 2011)
25. A. Roland, P. Shiman, W. Aspray, *Strategic Computing: DARPA and the Quest for Machine Intelligence, 1983–1993* (The MIT Press, 2002)
26. V. Kumar, P.S. Gopalakrishnan, L.N. Kanal, *Parallel Algorithms for Machine Intelligence and Vision* (Springer, 1990)
27. B. Meltzeb, D. Michie (eds.), *Machine Intelligence 7* (Edinburgh University Press, Edinburgh, 1972)
28. S. Legg, M. Hutter, *Universal Intelligence: A Definition of Machine Intelligence* (Springer, 2007)
29. J.R. Koza et al., *Genetic Programming IV: Routine Human-Competitive Machine Intelligence* (Springer, 2006)
30. N. Buduma, N. Locascio, *Fundamentals of Deep Learning: Designing Next-Generation Machine Intelligence Algorithms* (O'Reilly Media, 2017)

# Chapter 11
# Artificial Intelligence

## Contents



> AI is only as good as the humans programming it and the system in which it operates. If we are not careful, AI could not make healthcare better, but instead unintentionally exacerbate many of the worst aspects of our current healthcare system.
>
> —Bob Kocher

## 11.1 Introduction

Artificial Intelligence (AI) is a field of computer science that deals with intelligent machines, which are able to perform tasks heretofore only performed by human beings. It is mainly concerned with applying computers to tasks that require knowledge, perception, reasoning, understanding, and cognitive abilities. AI is potentially the algorithmic study of processes in every field of study [1]. AI deals with the intelligence demonstrated by machines, as opposed to natural intelligence displayed by humans. The main objective of AI is to teach the machines to think intelligently like humans do. It is to create technology that allows machines to function in an intelligent manner.

Artificial intelligence is a science with research activities in the areas of image processing, natural language processing, robotics, machine learning, etc. It began to


M. N. O. Sadiku, S. M. Musa, *A Primer on Multiple Intelligences*,
https://doi.org/10.1007/978-3-030-77584-1_11

become an active field of research within computer science in about 1955. AI has many applications in today's society. Although AI is a branch of computer science, there is hardly any field which is unaffected by this technology. Common areas of applications include agriculture, business, law enforcement, oil and gas, banking and finance, education, transportation, healthcare, automobiles, entertainment, manufacturing, speech and text recognition, facial analysis, and telecommunications. AI is redefining the job descriptions of all sectors of professions. Today industrial leaders such as Google, Microsoft, Procter & Gamble, and IBM have invested heavily on AI.

Artificial intelligence, with the capacity to make computers to learn from experience, is playing a predominant role in many industries. AI algorithms, such as machine learning and deep learning, are most commonly used to make intelligent predictions. AI has also received increased attention in recent years due to its applications in industrial real-time systems and investments in AI have grown over the years. An important feature of AI technology is that it can be added to existing technologies.

The chapter provides an introduction on artificial intelligence. It begins by explaining what artificial intelligence is all about. It addresses various components of AI. It discusses four major areas where AI technology can be applied, namely food industry, business, manufacturing, and medicine. It highlights the benefits and challenges of AI. It presents the future of AI. It concludes with some comments.

## 11.2 What Is Artificial Intelligence?

Artificial intelligence (AI) is one of the most important global issues of the twenty-first century. The modern field of artificial intelligence came into existence in 1955 when the term "artificial intelligence" (AI) was coined by John McCarthy. AI is the branch of computer science that deals with designing intelligent computer systems that mimic human intelligence (e.g., visual perception, speech recognition, decision-making, and language translation). The ability of machines to process natural language, to learn, to plan makes it possible for new tasks to be performed by intelligent systems. The main purpose of AI is to mimic the cognitive function of human beings and perform activities that would typically be performed by a human being. Without being taught by humans, machines use their own experience to solve a problem. Although the original vision for artificial intelligence was to simulate human intelligence, research effort has gradually shifted to autonomous systems that compete with people [2].

AI is stand-alone independent electronic entity that functions much like human expert. Today, AI is integrated into our daily lives in several forms, such as personal assistants, automated mass transportation, aviation, computer gaming, facial recognition at passport control, voice recognition on virtual assistants, driverless cars, and

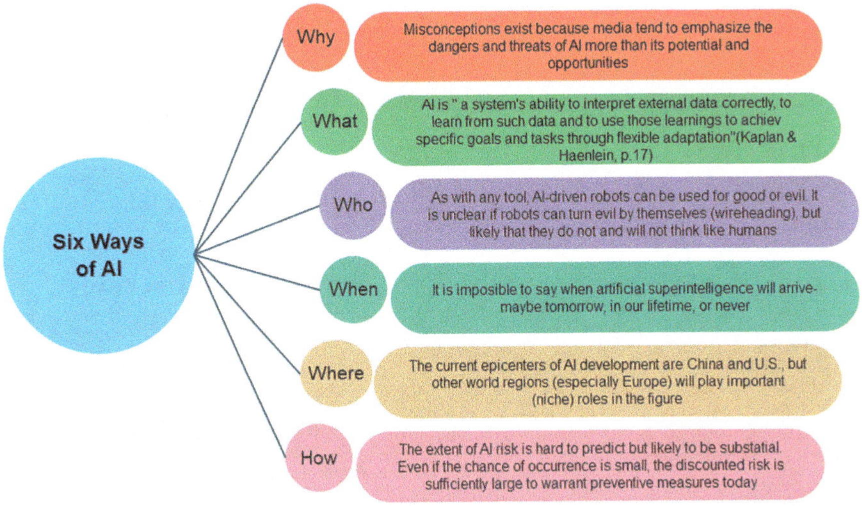


**Fig. 11.1** The 6 Ws of artificial intelligence

companion robots. AI technologies are performing better and better at analyzing data [3].

An important feature of AI technology is that it can be added to existing technologies. AI has benefited many areas such as chemistry and medicine, where routine diagnoses can be initiated by AI-aided computers. It embraces a wide range of disciplines such as computer science, engineering, chemistry, biology, physics, astronomy, neuroscience, and social sciences. Figure 11.1 illustrates the 6 Ws of artificial intelligence.

## 11.3 Major Components of AI

AI is not a single technology but a range of computational models and algorithms. AI techniques are pervasive and are too numerous to list here. The major components of AI include expert systems, fuzzy logic, and artificial neural networks (ANNs), machine learning, deep learning, natural language processing, computer vision, and robotics. These computer-based tools or technologies that have been used to achieve AI's goals are illustrated in Fig. 11.2 and explained as follows [4, 5]:

- *Expert Systems*: An expert system (ES) (or knowledge-based system) essentially embodies the knowledge and reasoning of human experts. An ES enables computers to make decisions by interpreting data and selecting between alternatives just as a human expert would do. It uses a technique known as rule-based inference in which rules are used to process data. The main business application

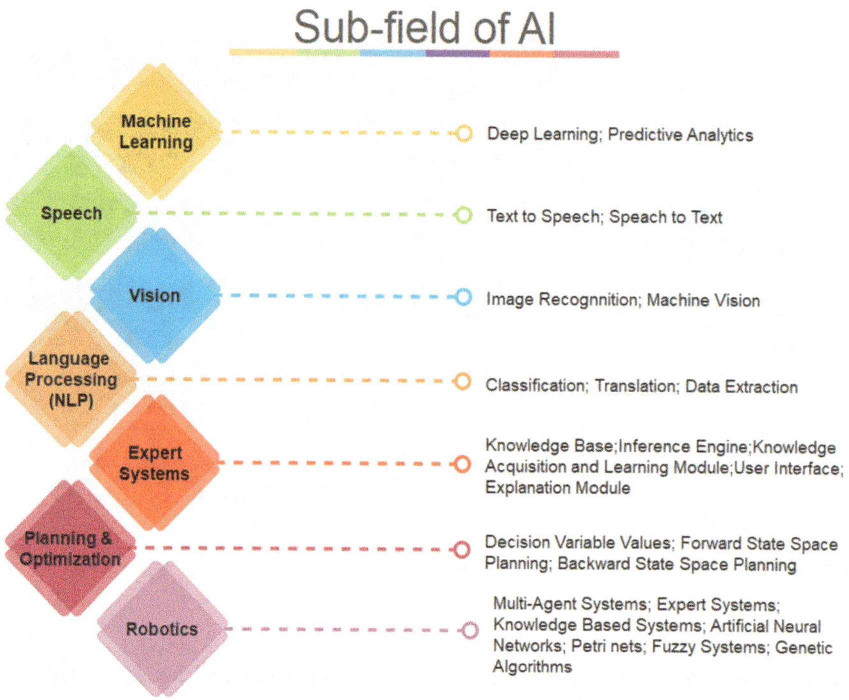


**Fig. 11.2** Subfields of artificial intelligence

of AI is in expert systems, which assist human experts in solving difficult problems.

- *Neural Networks*: These computer programs identify objects or recognize patterns after having been trained. Artificial neural networks (ANNs) are parallel distributed systems consisting of processing units (neurons) that calculate some mathematical functions. The ANN model represents nonlinear relationships which are directly learned from the data being modeled. Neural networks are being explored for healthcare applications in imaging and diagnoses, risk analysis, lifestyle management and monitoring, health information management, and virtual health assistance.
- *Computer Vision*: Computer vision comprehends methods and techniques through which artificial vision systems can be constructed. Essentially, a computer vision system is composed of image acquisition and image processing. Computer vision systems are already widely employed in different segments of agricultural production and industrial food production. They can be used in grading systems for orange, papaya, almond, potato, lemon, wheat, corn, rice, and soybean [6].
- *Natural Language Processors*: For AI to be useful to us humans, it needs to be able to communicate with us in our language. Computer programs can translate or interpret language as it is spoken by normal people. NLP techniques extract

information from unstructured data such as clinical notes to supplement and enrich structured medical data. NLP includes applications such as speech recognition, text analysis, translation, and other goals related to language. There are two basic approaches to NLP: statistical and semantic [7]. NLP allows for intelligent search engines, helpful chatbots, and accessibility for people who are visually impaired.

- *Robots*: These are computer-based programmable machines that have physical manipulators and sensors. Sensors can monitor temperature, humidity, pressure, time, record data, and make critical decisions in some cases. Robots have moved from science fiction to your local hospital. Today, robots perform vital functions in homes, industries, outer space, hospitals, and on military instillations. In jobs with repetitive and monotonous functions they might even completely replace humans. Robotics and autonomous systems are regarded as the fourth industrial revolution. Figure 11.3 shows a typical use of robots in agriculture [8].
- *Fuzzy Logic*: Reasoning based on imprecise or incomplete information in terms of a range of values rather than point estimates. Fuzzy logic deals with uncertainty in knowledge that simulates human reasoning in incomplete or fuzzy data. The fuzzy model is robust to parameter changes and tolerant to impression.
- *Machine Learning:* Algorithms make predictions and interpret data and "learn" without static program instructions. ML is a statistical technique for fitting models to data and training models with data. ML extracts features from input data by constructing analytical data algorithms and examines the features to create predictive models. The most common ML algorithms are supervised learning, unsupervised learning, reinforcement learning, and deep learning. ML algorithms are a good fit for anti-malware solutions because machine learning is well suited to solve "fuzzy" problems. Today, ML models are being developed in the food industry.
- *Deep Learning*: This is a subset of machine learning built on a deep hierarchy of layers, with each layer solving different pieces of a complex problem. It aims at increasing the capacity of supervised and unsupervised learning algorithms for

**Fig. 11.3** Use of robots in agriculture [8]

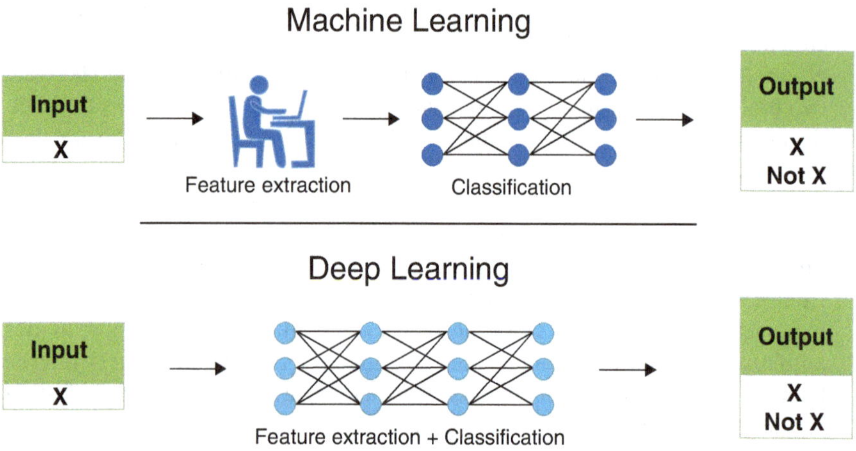


**Fig. 11.4** Difference between machine learning and deep learning

solving complex real-world problems by adding multiple processing [9]. The difference between machine learning and deep learning is shown in Fig. 11.4.

- *Data Mining:* This deals with the discovery of hidden patterns and new knowledge from large databases. Data mining exhibits a variety of algorithmic tools such as statistics, regression models, neural networks, fuzzy sets, and evolutionary models.

Each AI tool has its own advantages. Using a combination of these models, rather than a single model, is recommended. AI technologies are drastically influencing the retail industry and customer experience. Some types of artificial intelligence are predominant in business, while others are not. AI systems are designed to make decisions using real-time data. They have the ability to learn and adapt as they make decisions.

## 11.4 AI in Food Industry

The global food industry continues to grow steadily. The global population is increasing and is widening the gap between food supply and demand. Today, the global population exceeds seven billion. The rising global demand for food can only be met using technology. Several food companies have incorporated machine learning and deep learning into food and beverage products and services [10]. Key areas of applications of AI in food industry include production, product development, product customization, marketing, manufacturing, robotics, and processing of food products. AI can also be applied in restaurants, bar, and cafe businesses [11–14]. The following are common applications of AI in the food industry.

- *Agriculture:* Agriculture is facing major challenges and AI is regarded more and more as a solution to these challenges. There is an unquestionable growing tendency in the adoption of artificial intelligence in agriculture. Making the right decisions in agricultural activities is less than half the way needed to achieve desired results. The intelligent agents implementing measures into practice are seen more often as a feasible solution for the future. Agriculture automation is the main concern of every country. It has benefited from different automation practices like IoT, wireless communications, machine learning and artificial intelligence, and deep learning. However, the use of technologies in agriculture is still in infancy [15, 16].
- *Food Processing:* Food processing in many places is not fully automated. The food processing industry is benefiting from artificial intelligence, which is helping everything from sorting foods, maintaining health and safety compliances, developing new products, and improving the supply chain. The technology is basically helping to streamline work processes, making the work of employees easier, and making operations more efficient. Food processing industry can guarantee complete hygiene and high food quality by automating their processes as much as possible.
- *Food Manufacturing:* AI can be used by food companies in food manufacturing. Manufacturing of large productions of goods requires big, complicated, and intricately constructed mechanisms. AI in food manufacturing helps to monitor every stage of this process and makes predictions about cost and stock management. Machine learning allows identifying factors affecting the quality and causing flows in the manufacturing process and monitors the path of products from where they are produced and, ultimately, to where customers get it, ensuring transparency.
- *Food Production:* AI is used in advanced applications to make food production more efficient, safe, and profitable. It has enormous potential to optimize production and uncover manufacturing facilities' best operating points. It can enable faster production changeovers and identifying production bottlenecks before they become a problem. AI can optimize your facility's productivity, efficiency, and output.
- *Food Packaging:* AI-powered robotic equipment can accurately perform complex human tasks like the packaging. Robots are meeting the packing and picking demands accelerated by consumers' requests. The complex and labor-intensive nature of the process offers unique potential for intelligent automation.
- *Food Supply Chain*: The food industry needs to supply markets with high-quality food products at reasonable cost to satisfy consumer demands. Supply chains in the food industry are expanding steadily. To enhance food safety, companies can use AI's potential across their supply chain. The technology enables companies to test and monitor food safety products at every step of the supply chain. AI-powered supply chain management systems can be used to monitor and control the activities in the entire supply chain. AI can be used to minimize delays and maximize profit margins by providing close monitoring of every supply chain operation.

- *Food Delivery:* This is a growing segment of the food industry. The use of AI is a common enabler for most food delivery organizations. Leading food delivery services are using AI to enhance their marketing efforts through automation. AI applications also improve operational efficiency by automating activities like food ordering, dispatching, and billing processes. Food delivery service companies also use AI to enhance their customer services.
- *Restaurants*: AI is also being used by many apps to recommend the best restaurants based on the location, taste, and previous choices of the user. An AI in food service solution offers to unite the data from various food delivery programs. Self-serving systems are being massively taken up by restaurants as long as they enable customers to control the ordering process.
- *Robotics:* Robots are widely used in the food industry, but they are only for privileged big food businesses, still unavailable for small-to-medium. Robots are being used from seeding, spraying water and harvesting to cutting, processing and packaging of food products [17]. These are drones to deliver orders or robotic hands that can manage many processes in food manufacturing and even cooking. For example, 7-Eleven claims they use drones in their delivery service. Although robotics is still a quite subtle thing to introduce in some factories, it will bring an obvious benefit in the long run.

Other applications include food safety, food retailers, food market analysis, optical food sorting, predictive maintenance, tracking and traceability, personalized customer service, and heightened consumer engagement.

## 11.5 AI in Business

AI is being adopted today by businesses both big and small. AI use must be done with IT and the entire business working together. It does not generally replace human intelligence but serves as a supporting tool. AI seems destined to profoundly impact all aspects of business. It is widely used in business applications such as marketing, finance, accounting, human resources, supply chain, automation, data analytics, and natural language processing. Some of these are explained as follows [18–21]:

- *Marketing*: Marketing is one of the most notable areas of applications of AI. It may decide the next most effective marketing strategy. AI helps to develop marketing strategies and execute them. One way AI is marketing products is through chatbots, which can help solve problems, suggest products/services, and support sales. These AI chatbots can understand natural language. Retailers will be able to use AI technology to track shoppers in their stores and prevent theft. For example, Walmart has been using HANA to process high volume of transaction records.
- *Finance:* Investments in financial AI is growing as AI is increasingly being applied in financial institutions such as banks and insurance companies. Some

banks use various AI tools to detect fraudulent activity. Loan applications are now being processed by software that can take into account a variety of factors such as creditworthiness, credit score, and background check. Fraud detection represents another way AI is helpful in financial systems since AI can discern fraudulent activities. One of the first expert systems that helped with financial plans was created in 1986 by Applied Expert Systems (APEX) called the PlanPower.

- *Accounting:* Accounting is a discipline that deals with recording, analyzing, summarizing, and reporting every business transaction. Accounting can be divided into several parts based on the activity of the organization: financial accounting, cost accounting, and management accounting. The AI technology is revolutionizing the accounting field and altering the roles of accountants [22].
- *Automation:* The future is gravitating towards automation with AI driving the force behind eliminating the human error from business operations. AI can help improve efficiency and save money by automating many tasks. Automation alleviates repetitive or dangerous tasks. The increasing penetration of AI and autonomous devices into many aspects of life is improving efficiency and response times. For example, robotic process automation can produce amazing results in accounting department. The Apptus eSales solution is designed to automate merchandizing based on a predictive understanding of consumers. AI in automation will reduce errors and free up professionals from repetitive tasks. Business communication with customers can also be automated through online chats, email marketing, and social media.
- *Human Resources:* AI can help human resources (HR) departments by making candidate screening and recruitment process easier. Chatbots can be used to answer many commonly asked questions about company mission, policies, and benefits. Vendors in the HR sector such as Entelo, Textio, Textkernel, and HiringSolved offer AI solutions that help recruiters to sort and match potential workers and use bots to schedule candidate interviews. There is a notable impact from using AI in their staffing and talent management operations. AI can be a powerful tool in the hiring process.
- *Intelligent Supply Chain:* As consumer expectations continue to change, supply chains struggle to get the right products to customers, when and where they need them, without impacting margins. AI enables a self-learning and self-optimizing supply chain model with real-time insights to deliver precision.
- *Business Intelligence:* This may also be regarded as a collection of decision support technologies for the enterprise aimed at enabling professionals make better and faster decisions. The purpose of business intelligence (BI) is to support better business decision-making [23]. The use of AI and machine learning in BI is helping business enterprises to pull out actionable insights from complex data. Machine learning tools in business intelligence like the HANA will improve operational efficiency. BI chatbots are enabling decision-making by analyzing business data. AI has merged with BI applications in manufacturing and industrial sectors.

Other uses of AI in business include business analytics, chatbots and virtual assistance, fraud/crime detection, sales, R&D, e-commerce, advertising, customer service, talent selection and development, healthcare business, national security, criminal justice, logistics, transportation, and smart cities.

## 11.6 AI in Manufacturing

Manufacturing refers to the entire product life cycle: product design, production planning, production, distribution, and field service and reclamation [24]. Manufacturing is regarded as a prime generator of wealth and is critical in establishing a sound basis for economic growth. The leading manufacturing industries in the USA include steel, automobiles, chemicals, food processing, consumer goods, aerospace, and mining. Globalization has radically changed the manufacturing process in recent years. The manufacturing industry has always been receptive to adopting new technologies such as drones, industrial robots, AI, virtual reality, and Internet of things. These days, humans and robots collaborate to produce breakthroughs [25].

Artificial intelligence (AI) is one of the many technologies currently impacting manufacturing. Manufacturing efficiency and maintaining product quality remain top priorities in some industries such as IC industry, where process and equipment reliability directly influence cost, throughput, and yield [26]. Recent developments have enabled AI to cross into the mainstream: cloud computing, big data, machine learning, and Internet of things (IoT).

With automation, it is easy to believe that machines outperform human workers at every turn. It is possible for AI to develop a will of its own that may be in conflict with ours. Industrial robots perform what they were programmed to do, but they cannot make complex decisions as humans. The increasing use of connected technologies makes the manufacturing system vulnerable to cyberattacks [27]. Without doubt artificial intelligence holds the key to future growth and success in manufacturing. There is an excellent future for use of AI in manufacturing due to the potentially high value added when applied successfully [28].

The techniques of the artificial intelligence (AI) in general and machine learning (ML) in particular have been a source of inspiration for researchers in the manufacturing field, because ML enables computer programs automatically to improve their performance at some tasks through experience. The applications of ML techniques in manufacturing have increased over the last two decades due to many factors. Common manufacturing domains where ML is applied include the following.

- *Predictive Manufacturing:* The one of main objective of applying machine learning techniques to manufacturing systems is to accomplish the predictive manufacturing. This is a manufacturing system that has a self-awareness. It is the manufacturing system that can detect current conditions, diagnose a problem, and make a decision. This may also be regarded as predictive maintenance, which is able to predict disruptions to the production line in advance of that

disruption [29]. Data collected from sensors can be used to predict the failures and do preventive maintenance which is lot cheaper than fixing after failure. Predictive manufacturing is required in order to realize the smart factory.
- *Flexible Manufacturing:* Scheduling is well known as a part of any manufacturing system's control process. It assigns machines and other resources to jobs, or operations within jobs, in an efficient manner. Machine learning can be used to provide dynamic scheduling of flexible manufacturing system (FMS) [30]. For example, a multi-cell FMS consists of several independent cells, with each flexible manufacturing cell including several computer controlled machines, equipped with facilities for automating changing of parts and tools [31].
- *Smart Manufacturing:* This integrates big data, high-performance computing, and industrial Internet of things (IIoT) into traditional manufacturing systems to create highly customizable products with higher quality at lower costs. Industry 4.0 or smart manufacturing (or intelligent manufacturing) is required to be adapted through data information transfer between real world and machines. In the era of smart manufacturing, the abundance of smart sensors and industrial Internet of things has made manufacturing systems a data-rich environment.
- *Semiconductor Manufacturing*: Semiconductor manufacturing is a complex process that requires monitoring a great number of parameters from production to packaging. VLSI technology has advanced to the stage where the automation of semiconductor manufacturing has become imminent. Expert system technology can be applied to monitor various processes and equipment for the IC manufacturing environment. Machine learning is a useful tool in the automation of the IC manufacturing process. It also serves as an aid to engineers in interpreting and assimilating experimental results [32]. Within the semiconductor wafer manufacturing process, tight quality control is extremely important. Also, the volume of data in semiconductor manufacturing makes the data analysis difficult and time-consuming. Most semiconductor manufacturers keep a vast database to record every parameter in the manufacturing process. This online availability of vast amounts of production data brings the need for automatic analysis tools, to which machine learning techniques are good candidates [33].

Machine learning is also used in additive manufacturing, digital manufacturing, heterarchic manufacturing, and connected supply chain. Manufacturers are turning to machine learning to improve the end-to-end performance of their operations [34].

## 11.7 AI in Medicine

Artificial intelligence currently has several applications in healthcare and medicine. It has the potential to personalize medicine, work fluently in natural language, and handle big data. We can break down the applications in medicine into three separate categories: algorithmic solutions, visual tools, and medical practice. Today, AI is already being used in medicine in several areas such as decision support systems, laboratory information systems, robotic surgical systems, therapy, and reducing human error. The following are other common areas of applications.

- *Cancer:* The penetration of AI into cancer treatment affects treatment capabilities and safety. AI could be useful for head/neck cancers, prostate cancer, colorectal cancers, breast cancer, and cervical cancer. Machine learning algorithms include computer-aided detection applications that are used in mammography for the early diagnosis of breast cancer. AI will be part of the future of oncology [35].
- *Medical Imaging:* The number of imaging performed annually has skyrocketed over the last two decades. Medical imaging data provides a rich source of information about patients. AI can support radiologists and pathologists as they use medical imaging to diagnose a wide range of conditions, accelerate their productivity, and improve their accuracy. For example, when a patient complains about shortness of breath, the chest radiograph is likely the first imaging study to be conducted. AI could help improve accuracy and efficiency of polyp detection at CT colonography (CTC). AI-based computer-aided detection is routinely used in breast cancer screening programs [36].
- *Radiology*: A radiologist is a medical doctor who is skilled in interpreting medical images such as CT scans, MRI, digital radiographs, and ultrasounds. Today, radiologists are overwhelmed by the amount of imaging data they must review. AI technology, working alongside the radiologist, has the potential to improve accuracy, safety, efficiency, and productivity. In radiology, machine learning algorithms are being used to detect breast cancer, colonic polyps, and pulmonary nodules [37].
- *Aging:* AI can be used in aging and longevity research. Aging is a gradual, time-dependent process leading to the loss of function, biological and physical damage. It is a universal unifying feature possessed by all living organisms, tissues, and cells. Modern deep learning techniques can be used to develop age predictors. Most of the advances in deep learning in the context of aging research are in the area of biomarker development [38].

AI can be used to comb through mountains of health data (such as electronic health records [EHRs]), uncovering new insights and helping to predict disease risk. Other applications include clinical medicine, neurosurgery, dermatology, and ophthalmology, diabetes care, obstructive sleep apnea, cardiology, internal medicine, retina disease, drug development, vison and eye care, psychiatry, and hematology [39].

Machine learning techniques have been currently applied in the analysis of data in medicine. They were first applied to medicine with the use of electronic health records. They emerged in the medical sciences as clinical decision support techniques to improve sensitivity, disease detection, and monitoring. There are almost limitless opportunities for machine learning in medicine. Some applications of machine learning include [40, 41]:

- *Oncology:* Almost all works in this field apply machine learning techniques, which perform deep statistical analysis of a set of clinical cases supported by gene expression data. Researchers are using deep learning to train algorithms to recognize cancerous tissue at a level comparable to trained physicians. Google has developed a machine learning algorithm to help identify cancerous tumors on mammograms. Stanford is using a deep learning algorithm to identify skin cancer. Lately machine learning has been applied to cancer prognosis and prediction [42].
- *Pathology:* This is the medical specialty that is concerned with the diagnosis of disease. Machine vision and other machine learning technologies can enhance the efforts traditionally left only to pathologists with microscopes.
- *Genomic Medicine*: One of the goals of genomic medicine is to determine how variations in the DNA of individuals can affect the risk of different diseases. Machine learning can help to model the relationship between DNA and the quantities of key molecules in the cell [43].
- *Predicting Chronic Disease:* Machine learning can help hospital systems identify patients with chronic disease, predict the likelihood that patients will develop chronic disease, and present patient-specific prevention interventions. For example, ML approaches have significant potential in prediction of acute exacerbations in asthma patients [44].
- *Personalized Medicine:* This is individualization, recognition of the micro-variables within a patient that may cause them to be different from their peers. ML for personalized medicine is a growing area. It involves the ability to draw on large data sets and predictive models, which allow for clinicians to more confidently diagnose, predict, and treat their patients.
- *Reduce Readmissions*: Machine learning can reduce readmissions. Clinicians can receive daily guidance as to which patients are most likely to be readmitted and how they might be able to reduce that risk.

These advances would have been unimaginable without machine learning. Other areas include surgery, radiology, emergency medicine, traditional medicine, mental health, psychiatry, ophthalmology, and optometry. ML algorithms have immense potential to enhance diagnostic and intervention research in smoking, depression, asthma, and chronic obstructive pulmonary disease (COPD). ML is also being used for data analysis and interpretation of continuous data used in the Intensive Care Unit [45]. With time, the healthcare environment will become more and more reliant on computer technology and ML capabilities will reach into all aspects of medicine.

Other areas of applications of AI include chemical industry, hospitality industry, cybersecurity, banking, social media, security, criminal justice, space exploration, military, creative art, music, automotive, aviation, agriculture, law enforcement, oil and gas, banking and finance, education, transportation, automobiles, entertainment, image processing, speech and text recognition, facial analysis, and telecommunications.

## 11.8 Benefits

Artificial intelligence can enhance and extend the capabilities of humans. AI systems deliver more accurate production lines results with greater speed and more consistency than human workers. AI solutions display great potential for the optimization of the hygiene and cleaning tasks, which is the factor number one to influence food safety. Businesses can implement intelligent algorithms to improve their quality of food and services which in turn will mean many healthy meals for the customers [46]. AI could utilize vast data sets of detailed agricultural information for the improvement of our food crops, faster than ever before. Adoption of AI can position companies as digital leaders.

Companies of all types and sizes are finding ways to use the right artificial intelligence technology to save time and money. In order to meet evolving customer expectations, companies are leveraging intelligent technologies to transform the way they operate.

## 11.9 Challenges

Just like other technologies, AI comes with challenges, such as accountability, security, technological mistrust, and the displacement of human workers. Widespread use of artificial intelligence could have unintended, undesirable consequences. AI is also fueling anxieties and ethical concerns, raising the issue of trustworthiness of AI systems, and infringing on human rights and values, such as privacy. These are clear challenges that must be addressed to support AI technology's future. The stakeholders must ensure that AI's impact is a positive one by proactively tackling the challenges, while ensuring the opportunities remain available.

AI is yet to be ubiquitous due to cost constraints and lack of skilled experts. The high cost of large-scale deployment in the sector is restricting the market to grow. Integrating new technologies, such as the integration of AI in food companies, is never easy [47]. Besides the USA and Europe, the adoption of AI is still in the early stages in many nations.

The field of AI is broad and even confusing, presenting a challenge and a barrier hindering wide application. A major challenge is the ongoing trend of the manufacturing domain to becoming more complex and dynamic. AI will never replace humans in the food industry, as humans will always be needed to oversee operations, repair, and maintain old equipment. The technology can essentially work side-by-side with humans to increase operational efficiency. AI will encounter more challenges in future development.

## 11.10 The Future of AI

Artificial intelligence is committed to making machines intelligent. The main advances of AI have been in search algorithms, machine learning algorithms, and integrating statistical analysis into understanding the world at large. The current popularization of the Internet, universal existence of sensors, emergence of big data, and other emerging technologies have caused rapid advancement in AI, leading to a new Artificial Intelligence 2.0 [48]. Figure 11.5 shows the evolution of artificial intelligence.

There are three types of artificial intelligence [49, 50]:

- *Artificial Narrow Intelligence* (ANI): Artificial intelligence today is properly known as narrow AI (or weak AI), in that it is designed to perform a narrow task (e.g., only facial recognition). In other words, these AI systems are designed to execute a single task really well.
- *Artificial General Intelligence* (AGI): Future AI will be known as general AI (AGI or strong AI). While narrow AI may outperform humans at whatever its specific task is, AGI would outperform humans at nearly all cognitive tasks. This AI system has a human level of cognitive functions such as language processing, image processing, and reasoning capabilities. We are far away from building an AGI system. The next 10–25 years will be characterized by a merger of narrow-AI and strong-AI approaches.
- *Artificial Super Intelligence* (ASI): By definition, a superintelligent AI is good at attaining its goals, so we need to ensure that its goals and our goals are properly aligned. Superintelligent AI will help us develop the machinery which will be able to understand human completely. An ASI system would be able to surpass all human capabilities. An ASI has the capability of modifying itself into new forms and presenting an existential risk to the human species. Once humans develop ASI, it will take off on its own and redesign itself at an ever-increasing

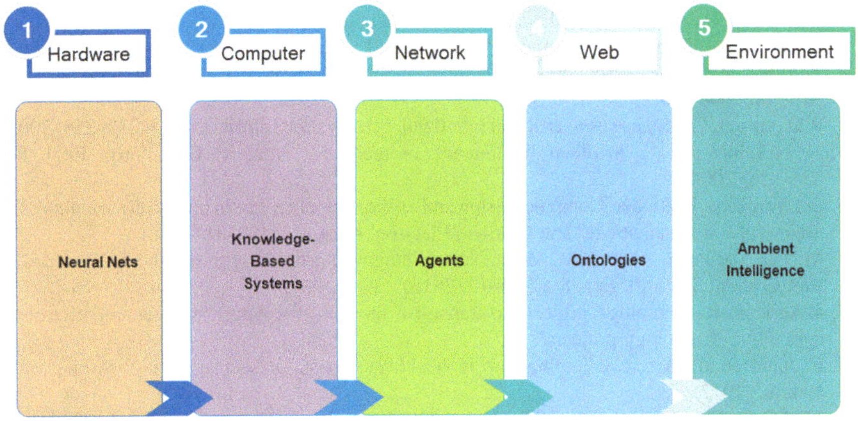


**Fig. 11.5** The evolution of artificial intelligence

rate. Some are of the opinion that superhuman AI will never arrive because it is physically impossible.

## 11.11 Conclusion

Artificial Intelligence (AI) is a way of making intelligent machines that think, work, and react like humans. It is the theory and development of computer systems that can perform tasks that normally require human intelligence. It deals with computational systems that imitate the intelligent behavior of expertise.

AI technology can change the world for the better. AI is pervading and transforming all aspects of the food industry, manufacturing industry, chemical industry, medicine/healthcare, business, etc. While there is much promise in AI, the future depends on deeper engagement among specific industry, policymakers, and technologists. Tech giants like Google, Microsoft, and IBM are highly interested in developing AI. The entire society can benefit from the use of AI. It is time to address world hunger using AI and machine learning.

Several academic institutions and companies have started to offer courses in artificial intelligence. More information on artificial intelligence can be found in the books in [51–62] and the following related journals:

- *Artificial Intelligence Review*
- *Artificial Intelligence in Agriculture*
- *Journal: Applied Artificial Intelligence*
- *Artificial Intelligence for Engineering Design, Analysis and Manufacturings*

## References

1. M.N.O. Sadiku, Artificial intelligence. IEEE Potent., 35–39 (1989)
2. Artificial intelligence. Commun. ACM **60**(2), 10–11 (2017)
3. Y. Mintz, R. Brodie, Introduction to artificial intelligence in medicine. Minim. Invasive Ther. Allied Technol. **28**(2), 73–81 (2019)
4. R.O. Mason, Ethical issues in artificial intelligence. Encyclop. Inform. Syst. **2**, 239–258 (2003)
5. A.N. Rames et al., Artificial intelligence in medicine. Ann. R. Coll. Surg. Engl. **86**, 334–338 (2004)
6. D.I. Patrícioa, R. Rieder, Computer vision and artificial intelligence in precision agriculture for grain crops: a systematic review. Comput. Electron. Agric. **153**, 69–81 (2018)
7. M.N.O. Sadiku, Y. Zhou, S.M. Musa, Natural language processing in healthcare. Int. J. Adv. Res. Comput.Sci. Soft. Eng. **8**(5), 39–42 (2018)
8. 4 Agriculture technology projects to transform the industry, https://www.ennomotive.com/agriculture-technology-projects/
9. F. Jiang et al., Artificial intelligence in healthcare: past, present ad future. Stroke Vasc. Neurol. (2017)
10. M.N.O. Sadiku, T.J. Ashaolu, S.M. Musa, Emerging technologies in agriculture. Int. J. Scient. Adv. **1**(1), 31–34 (2020)

11. K. Utermohlen, 4 Applications of artificial intelligence in the food industry (2019), https://heartbeat.fritz.ai/4-applications-of-artificial-intelligence-ai-in-the-food-industry-e742d7c02948#:~:text=1)%20Sorting%20Food,of%20canned%20and%20bagged%20goods.&text=One%20of%20the%20most%20advanced,solutions%20with%20machine%20learning%20functionalities
12. Machine learning and artificial intelligence in the food industry (2019), https://medium.com/@spd.group/machine-learning-and-artificial-intelligence-in-the-food-industry-598f78471106
13. How artificial intelligence is revolutionizing the food and beverage industry (2020), https://new.abb.com/news/detail/61772/how-artificial-intelligence-is-revolutionizing-the-food-and-beverage-industry#:~:text=AI%20can%20play%20a%20vital,complexity%20to%20the%20production%20line
14. Applications of artificial intelligence (AI) in the food industry, https://morioh.com/p/6f346ea78169
15. Adoption of artificial intelligence in agriculture. Bull. UASVM Agricult. **68**(1), 284–293 (2011)
16. K. Jha et al., A comprehensive review on automation in agriculture using artificial intelligence. Artific. Intell. Agricult. **2**, 1–12 (2019)
17. J. Iqbal, Z.H. Khan, A. Khalid, Prospects of robotics in food industry. Food Sci. Technol. **37**(2), 159–165 (2017)
18. Business applications for artificial intelligence: an update for 2020 (2019), https://blog.dce.harvard.edu/professional-development/business-applications-artificial-intelligence-what-know-2019
19. B. Marr, 10 Business functions that are ready to use artificial intelligence (2020), https://www.forbes.com/sites/bernardmarr/2020/03/30/10-business-functions-that-are-ready-to-use-artificialintelligence/#70fe2c923068
20. D.M. West, J.R. Allen, How artificial intelligence is transforming the world (2018), https://www.brookings.edu/research/how-artificial-intelligence-is-transforming-the-world/
21. Why is artificial intelligence in business analytics so critical for business growth? (2019), https://medium.com/gobeyond-ai/why-is-artificial-intelligence-in-business-analytics-so-critical-for-business-growth2ed16cbe7846#:~:text=AI%2Dpowered%20BI%20tools%20have,day%2Dto%2Dday%20decisions
22. D. Kwafo, The impacts of artificial intelligence on management accounting students: a case study at Oulu Business School, University of Oulu, Master's Thesis, University of Oulu, 2019
23. M.N.O. Sadiku, M. Tembely, S.M. Musa, Toward better understanding of business intelligence. J. Scient. Eng. Res. **3**(5), 89–91 (2016)
24. M.S. Fox, Industrial applications of artificial intelligence. Robotics **2**, 301–311 (1986)
25. M.N.O. Sadiku, S.M. Musa, O.M. Musa, Artificial intelligence in the manufacturing industry. Int. J. Adv. Scient. Res. Eng. **5**(6), 108–111 (2019)
26. Artificial intelligence in semiconductor manufacturing, in *Wiley Encyclopedia of Electrical and Electronics Engineering*, ed. by J. Webster, (Wiley, 2007)
27. J. Lee et al., Industrial artificial intelligence for Industry 4.0-based manufacturing systems. Manufact. Lett. **18**, 20–23 (2018)
28. T.M. Knasel, Artificial intelligence in manufacturing: forecasts for the use of artificial intelligence in the USA. Robotics **2**, 357–362 (1986)
29. J.H. Han, R. Kim, S.Y. Chi, Applications of machine learning algorithms to predictive manufacturing: trends and application of tool wear compensation parameter recommendation, in *Proceedings of the 2015 International Conference on Big Data Applications and Services*, Jeju Island, Korea (2015), pp. 51–57
30. P. Priore et al., A review of machine learning in dynamic scheduling of flexible manufacturing systems. Artif. Intell. Eng. Design Anal. Manufact. **15**, 251–263 (2001)
31. Y. Arzi, A. Herbon, Machine learning based adaptive production control for a multi-cell flexible manufacturing system operating in a random environment. Int. J. Prod. Res. **38**(1), 161–185 (2000)
32. K.B. Irani et al., Application of machine learning techniques to semiconductor manufacturing, in *Proceedings of the SPIE 1293, Applications of Artificial Intelligence VIII* (1990)

33. C.K. Shin, S.C. Park, A machine learning approach to yield management in semiconductor manufacturing. Int. J. Prod. Res. **38**(17), 4261–4271 (2000)
34. M.N.O. Sadiku, Y.P. Akhare, S.M. Musa, Machine learning in manufacturing: a brief survey. Int. J. Trend Res. Develop. **6**(6) (2019)
35. R.F. Thompson et al., The future of artificial intelligence in radiation oncology. Int. J. Radiat. Oncol. Biol. Phys. **102**(2), 247–248 (2018)
36. M.I. Faza et al., The past, present and future role of artificial intelligence in imaging. Eur. J. Radiol. **105**, 246–250 (2018)
37. T. Nawrocki et al., Artificial intelligence and radiology: have rumors of the radiologist's demise been greatly exaggerated? Acad. Radiol. **25**(8), 967–972 (2018)
38. A. Zhavoronko et al., Artificial intelligence for aging and longevity research: recent advances and perspectives. Ageing Res. Rev. **49**, 49–66 (2019)
39. M.N.O. Sadiku, T.J. Ashaolu, S.M. Musa, Artificial intelligence in medicine: a primer. Int. J. Trend Res. Develop. **6**(1), 270–272 (2019)
40. K. Sennaar, Machine learning for medical diagnostics – 4 current applications (2018), https://emerj.com/ai-sector-overviews/machine-learning-medical-diagnostics-4-current-applications/
41. G.S. Handelman, eDoctor: machine learning and the future of medicine. J. Intern. Med. **284**(6), 603–619 (2018)
42. J.A. Cruz, D.S. Wishart, Applications of machine learning in cancer prediction and prognosis. Cancer Informat. **2**, 59–78 (2006)
43. M.K.K. Leung et al., Machine learning in genomic medicine: a review of computational problems and data set. Proc. IEEE **104**(1), 176–197 (2016)
44. J. Finkelstein, I.C. Jeong, Machine learning approaches to personalize early prediction of asthma exacerbations. Ann. N. Y. Acad. Sci. **1387**, 153–165 (2017)
45. G.D. Magoulas, A. Prentza, Machine learning in medical applications, in *Advanced Course on Artificial Intelligence*, ed. by G. Paliouras, V. Karkaletsis, C. D. Spyropoulos, (Springer, 1999), pp. 300–307
46. A. Eliazàt, 4 Ways AI is revolutionizing the food industry (2020), https://autome.me/4-ways-ai-is-revolutionizing-the-food-industry/
47. Artificial intelligence applications in food industry (2019), https://www.linkedin.com/pulse/artificial-intelligence-applications-food-industry-mert-damlapinar
48. Y. Pan, Heading toward artificial intelligence 2.0. Engineering **2**, 409–413 (2016)
49. M. Tegmark, Benefits and risks of artificial intelligence, https://futureoflife.org/background/benefits-risks-of-artificial-intelligence/
50. V. Advani, What is artificial intelligence? How does AI work, types and future of it? (2020), https://www.mygreatlearning.com/blog/what-is-artificial-intelligence/
51. R.S. Mitchell, J.G. Michalski, T.M. Carbonell, *An Artificial Intelligence Approach* (Springer, 2013)
52. I. Rahwan, G.R. Simari, *Argumentation In Artificial Intelligence* (Springer, 2009)
53. A.H. Bond, L. Gasser, *Readings in Distributed Artificial Intelligence* (Morgan Kaufmann, 2014)
54. C.J. Simon, *Will Computers Revolt?: Preparing for the Future of Artificial Intelligence* (Future AI, 2018)
55. W. Ertel, N.T. Black, *Introduction to Artificial Intelligence*, 2nd edn. (Springer, 2018)
56. K.R. Chowdhary, *Fundamentals of Artificial Intelligence* (Springer, 2020)
57. K. Frankish, W.M. Ramsey, *The Cambridge Handbook of Artificial Intelligence* (Cambridge University Press, 2014)
58. K. Hammond, *Practical Artificial Intelligence for Dummies: Narrative Science Edition* (Wiley, Hoboken, NJ, 2015)
59. M. Yao, A. Zhou, M. Jia, *Applied Artificial Intelligence: A Handbook for Business Leaders* (TOPBOTS, 2018)
60. T. Wischmeyer, T. Rademacher (eds.), *Regulating Artificial Intelligence* (Springer, 2020)
61. D.W. Patterson, *Introduction to Artificial Intelligence* (Pearson Education India, 2015)
62. N.J. Nilsson, *Principles of Artificial Intelligence* (Morgan Kaufmann, 1982)

# Chapter 12
# Computational Intelligence

**Contents**



Computations are everywhere, once you begin to look at things in a certain way.

—Rudy Rucker

## 12.1 Introduction

Computational intelligence (CI) refers to recreating human-like intelligence in a computing machine. It consists of a set of computing systems with the ability to learn and deal with new situations such that the systems are perceived to have some attributes of intelligence. CI techniques are computational systems that try to mimic human behavior, such as perception, reasoning, learning, evolution, and adaptation. It is concerned with discovery of structures in data and recognition of patterns. It involves using an algorithm to capture hidden knowledge from data and use it to train an "intelligent machine" to make complex decisions without human intervention [1]. It encompasses techniques such as neural networks, fuzzy logic, and evolutionary computation which derive rules, patterns, and develop complex mappings from data.

The concept of computational intelligence (soft computing) was first used in 1990 by the IEEE Neural Networks Council, which later became the IEEE Computational Intelligence Society. CI provides knowledge engineers with a set of robust techniques and intelligent tools for various applications. CI methods provide the ability to learn from experience and extract knowledge.



M. N. O. Sadiku, S. M. Musa, *A Primer on Multiple Intelligences*, https://doi.org/10.1007/978-3-030-77584-1_12

Some consider CI as a branch of AI, while others consider them as parallel disciplines. Although Artificial Intelligence (AI) and Computational Intelligence (CI) seek a similar long-term goal, they are different. While AI is based on hard computing methods, CI is based on soft computing methods, which enable adaptation to many situations. One may regard AI as human-based, and CI as nature-based [2].

This chapter provides a brief introduction to computational intelligence. It begins by discussing what computational intelligence is all about. It addresses common computational intelligence techniques. It presents some typical applications of CI. It highlights the main benefits and challenges of CI. The last section concludes with comments.

## 12.2 What Is Computational Intelligence?

Computational intelligence (CI) refers to the ability of a computer to learn a specific task from experimental observation. It is an emerging discipline which provides powerful tools for modeling and analyzing complex systems. Some regard CI as a branch of artificial intelligence that comprises algorithms inspired by nature to address complex real-world problem. CI is the study of the design of intelligence systems. A system is regarded as "intelligent" only if it satisfies learning and decision-making requirements. It is familiar that the best-known manifestation of intelligence is human intelligence. The characteristic of "intelligence" is usually attributed to humans so that CI is a way of performing like human beings and using human-like reasoning, i.e., it uses inexact and fuzzy knowledge. Thus the goal of CI is to recreate human-like intelligence in a human-made machine.

The term "computational intelligence" was coined by John McCarthy in 1956. CI consists of a set of heuristics which exhibit intelligence such as learning, improvement through learning, flexibility, and adaptation. The ongoing worldwide computerization has created new opportunities for researchers. All branches of science and art have become computational: computational biology, computational physics, computational chemistry, computational ecology, computational linguistics, computation electromagnetic, computational finance, computational mechanics, computational social science, computational epistemology, computational intelligence, and so on. Computational intelligence is known for including powerful techniques such as artificial neural networks, fuzzy systems, evolutionary computation, learning theory, or probabilistic methods.

Although Artificial Intelligence (AI) and Computational Intelligence (CI) seek a similar goal of developing intelligent machines, they are different. AI is based on hard computing techniques, while CI is based on soft computing methods. AI operates in a top-down manner, while CI operates in a bottom-up manner. Some regard CI to be a subset or branch of AI [3]. Some of the most successful AI systems are based on CI. Hard (traditional) computing emphasizes precision, certainty, and rigor, while soft computing requires that computation, reasoning, and decision-making exploit the tolerance for imprecision and uncertainty wherever possible.

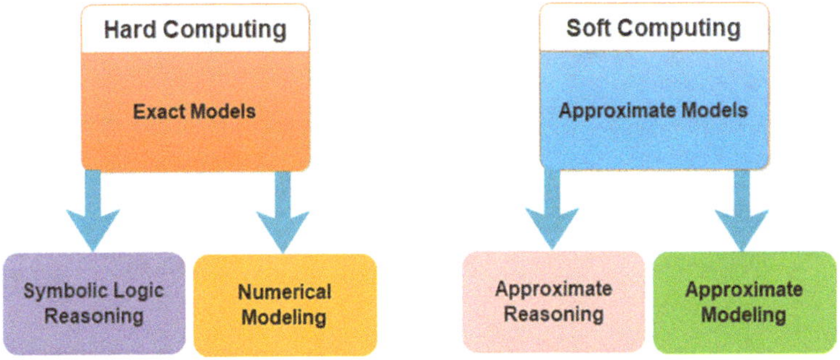


**Fig. 12.1** Hard and soft computing

**Fig. 12.2** Relationship between AI, CI, and soft computing

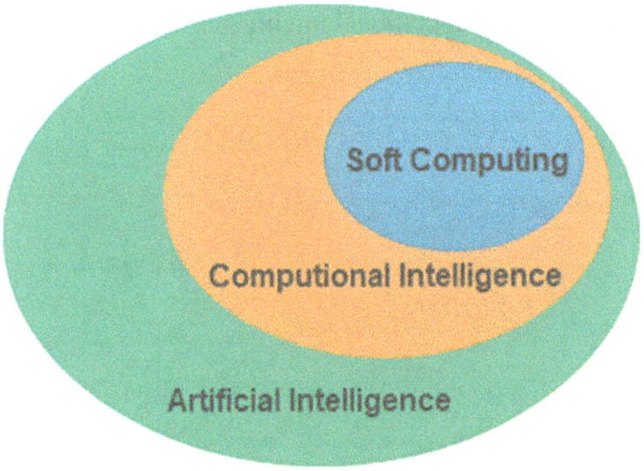


Figure 12.1 illustrates hard and soft computing. The relationship between AI, CI, and soft computing is depicted in Fig. 12.2.

## 12.3 Computational Intelligence Techniques

CI approaches aim at creating intelligent systems by using innovative and nature-inspired algorithms. Traditionally CI therefore uses a combination of five main complementary techniques: Neural Networks, Fuzzy Systems, Evolutionary Computation, Genetic Algorithm, and Probabilistic Reasoning. Some of these CI techniques have their origins in biological systems and are shown in Fig. 12.3. They are briefly explained as follows [4–7].

- *Neural Networks (NNs)*: These are the most popular artificial learning tools. Artificial neural network (ANN) is a CI method that mimics the neural system of human brain. NNs are parallel distributed networks that have the ability to learn and generalize from examples. NNs include feedforward NNs, recurrent NNs, self-organizing NNs, deep learning, and convolutional neural networks. An NN architecture typically consists of an input layer, one or more hidden layers, and

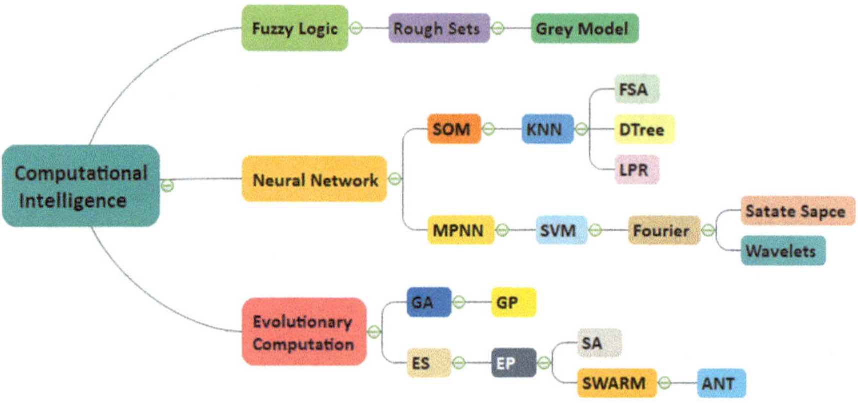


**Fig. 12.3** Computational intelligence techniques

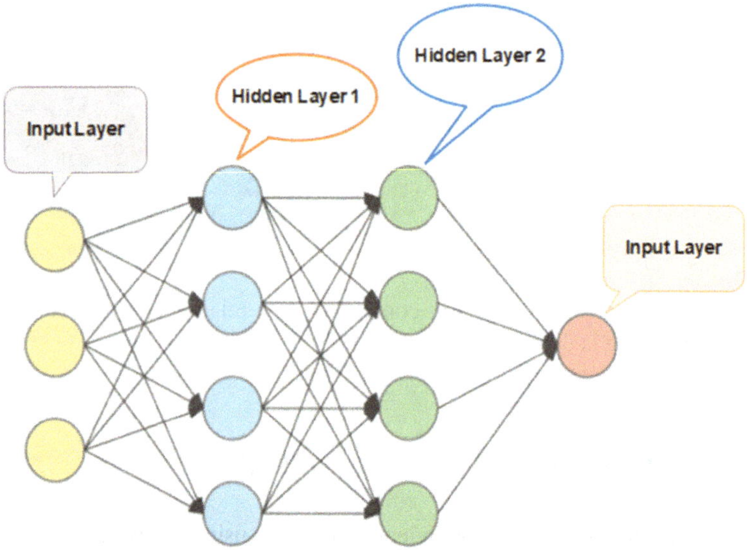


**Fig. 12.4** Typical structure of ANN

an output layer. With large numbers of input variables, it can be difficult for the user to understand the logic represented in this mapping of inputs to output. A typical structure of ANN is shown in Fig. 12.4.

- *Fuzzy Systems*: Fuzziness refers to the inexact or imprecise nature of common terms. It constitutes a form of "approximate reasoning." This area of research includes fuzzy sets and systems, fuzzy clustering and classification, fuzzy controllers, linguistic summarization, and fuzzy neural networks. Fuzzy algorithms attempt to capture the uncertainty and imprecision in rule-based representations that are not easily quantified by other methods. They are very powerful for modeling systems where the descriptors or inputs do not cleanly separate into discrete values or are subjective.

- *Evolutionary Computation*: Evolutionary computation (EC) is a CI model used to mimic the biological evolution phenomenon. EC is a computational intelligence technique inspired from natural evolution. It solves optimization problems by generating, evaluating, and modifying a population of possible solutions. EC currently includes five algorithms: genetic algorithm (GA), evolutionary programming (EP), evolution strategies (ES), genetic programming (GP), and swarm intelligence (SI). EC is very useful for optimizing the connections and weightings between the input layer, the hidden layer(s), and the output layers for an NN model.
- *Genetic Algorithm*: This is a subset of evolutionary computation. It takes all its inspiration from nature. The concept of genetic algorithm is to mimic the natural selection in nature where the fittest elements are chosen.
- *Probabilistic Reasoning*: Probabilistic reasoning is often used in dealing with uncertain and incomplete data in artificial intelligence. The aim of probabilistic reasoning is to combine the capability of probability theory with the capability of deductive logic to exploit structure.

Today, these computing techniques are used in classification, regression, pattern recognition, medicine, system modeling, decision-making, pattern, and food engineering. CI is an evolving field. In addition to the five main constituents, it encompasses computing paradigms like ambient intelligence, swarm intelligence, artificial immune systems, cultural learning, data mining, natural language processing, and artificial intelligence. The common characteristics of these techniques are their collective intelligence and adaptability to a changing environment. Due to their efficiency and simplicity, the algorithms have been employed successful for problem-solving across social and natural sciences [8].

## 12.4 Applications

CI plays a major role in developing successful intelligent systems, including games and cognitive developmental systems. It provides solutions to complex real-world problems to which mathematical or traditional modeling does not perform well. Researchers have used the CI techniques to design intelligent and context-aware algorithms to incorporate the aforementioned features. Successful applications of CI have been found in computer science, engineering, data analysis, business, marketing, medical, e-government, and manufacturing domains.

- *Wireless sensor networks* (WSNs): These are networks of distributed autonomous devices that can sense or monitor environmental conditions. WSNs face many challenges, mainly caused by communication failures, storage and computational constraints, and limited power supply. Computational intelligence has been successfully applied to address various challenges such as data aggregation and fusion, energy aware routing, task scheduling, security, optimal deployment, and localization. CI provides adaptive mechanisms that exhibit intelligent behavior in complex and dynamic environments like WSNs [9].

- *Business:* Computational intelligence techniques have been used to develop business intelligence and solve financial market problems. The techniques have been applied to many areas in business such as marketing, e-commerce, financial management, decision-making, classification, time series forecasting, and optimization [10]. CI simplifies the problem of predicting future financial markets.
- *Finance and Economics:* The financial sector is crucial to the economical and social organization of modern society. The number of applications of CI in this area is huge and could cover anything from asset pricing or game analysis to time series forecasting. The range of approaches that have already used in economics and finance includes fuzzy systems, neural networks, evolutionary computation, support vector machines, and ant algorithms. The techniques have been used in financial economics, games, industrial organization, macroeconomics, and econometrics [11].
- *Control:* CI methodologies are currently applied to all areas of control. Application of CI in control leads to neural network control, fuzzy control, reinforcement learning, and brain machine interfaces. Neural network has been successfully applied to system control with its universal approximation and adaptation capabilities [12].
- *Medicine*: Computational intelligence can be useful for medical purposes. CI models can analyze large amounts of data in order to predict healthcare outcomes for individual patients. They can handle the imprecision and uncertainty which is typically apparent in clinical and biological data. They are suitable for clinical predictions as they are efficient in finding patterns in data which contain noisy features, and which traditional statistical models alone often fail to handle [13]. Researchers have used various CI techniques for early diagnosis of heart disease and to improve the heart disease diagnosis accuracy [14]. The application of CI can be either diagnostic or therapeutic. It can help diagnose diseases and in preventive purpose [15]. Applying CI techniques in complex medical data can provide better management, faster performance, and higher level of accuracy.
- *Music Composition:* CI techniques such as neural networks, fuzzy systems, and evolutionary computation provide powerful tools for modeling, learning, and optimization. These techniques have been applied successfully to music composition, which involves musical elements such as timbre, pitch, rhythm, motive, phrase, and chord. The main task of music composition involves deliberating the musical form, creating the melody, and arranging the accompaniment [16].
- *Social creativity*: CI can be used to leverage social creativity. Creative solutions to complex problems faced by organization come from joint thinking as opposed to traditional views of creativity that focus on the individual as the major contributor of creativity. CI can broaden group participation by linking groups with

social networks and can facilitate divergent thinking and inspiration by linking users to huge amounts of online information and social media [17].

Other applications of CI include production electronics, games, hydrology, drug discovery and development, HVAC (Heating Ventilating and Air Conditioning) systems, transfer learning, sports, bioinformatics, intrusion detection, information systems, biological systems, power systems (smart grid), oil and gas reservoir, air pollution monitoring, manufacturing systems, corrosion damage assessments, coral reef ecosystems, environmental planning, logistics and supply chain management, etc. The combination of big data and computational intelligence provides a powerful tool. CI has the potential to influence modern societies, businesses, and the public [18].

## 12.5 Benefits

Computational intelligence is a set of nature-inspired computational techniques for addressing complex real-world problems for which traditional modeling does not perform well. It is efficient in solving real-world problems which require reasoning and decision-making. It produces more robust, simpler, and tractable solutions than the traditional techniques. A long-term goal for CI is creating cognitive systems that could compete with humans in large number of areas.

CI techniques can solve real-world complex problems in which traditional approaches are ineffective. They perform efficient approximations within reasonable computation time because they often use non-deterministic techniques. Nonlinear characteristics of the optimization models and 3D problems pose challenge for traditional mathematical programming methods. CI techniques perform well in nonlinear optimization problems and allow modeling systems of different complexity [19]. Other advantages of CI methods are that they often work efficiently, that they are fault tolerant, and that they are mostly well suited for parallelization.

## 12.6 Challenges

The biggest disadvantage of CI, compared to deterministic methods, is that the optimal solution(s) can generally be neither proved nor guaranteed to be found [20]. Attempt to combine different CI methods is a challenge. Another main challenge facing computational intelligence approaches is the problem of extracting knowledge from data.

Each of the various CI techniques has some advantages and suffers from some disadvantages. Therefore, there is no solution methodology that is universally superior. Hybrid methods may generate better results than single methods.

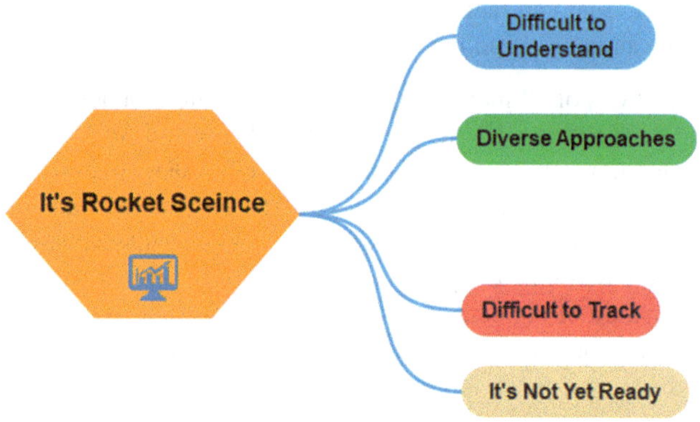


**Fig. 12.5** Some key issues in computational intelligence

There are those who believe that CI is a dead-end, unachievable attempt to recreate intelligence in a computing machine. This belief is based on inconsistent and contradictory goal of CI—while "computational" implies data processing ability, "intelligence" implies the ability to process information [21].

There has been some issues like security and privacy in data analysis of some sensitive information such as in medical applications. There are also challenges of big data procured from complex healthcare sources [22]. Some of the key issues facing computational intelligence are presented in Fig. 12.5.

## 12.7 Conclusion

Computational intelligence (CI) is a newly evolving research field in artificial intelligence. It uses various techniques such as artificial neutral network, evolutionary computation, fuzzy system, support vector machine, deep learning, swarm intelligence, and hybrid methods. The major goal of researchers in CI is mimicking Nature with artificial technologies. CI develops intelligent system with the ability to learn, deal with new situations, and reason. It is giving birth to blended intelligence and web intelligence.

The growing interest in CI research can perhaps be best illustrated by the increasing amount of journals, books, and articles published on CI in the recent years. As more and more engineering applications use CI techniques, the need to train students in this area increases [23]. CI is yet to reach maturity as a discipline in its own right. For more information on computational intelligence, see the books in [20, 24–55] and several books on Amazon.com. A lot of information is also available in the following journals devoted to CI:

- *Computational Intelligence*
- *Computational Intelligence. An International Journal*

- *Computational Intelligence and Neuroscience*
- *Applied Computational Intelligence and Soft Computing*
- *IEEE Transactions on Emerging Topics in Computational Intelligence*
- *IEEE Transactions on Computational Intelligence and AI in Games*
- *IEEE Computational Intelligence Magazine*
- *Journal of Computational Intelligence in Finance*
- *Journal of Computational Intelligence in Bioinformatics*
- *Journal of Advanced Computational Intelligence and Intelligent Informatics*
- *International Journal of Computational Intelligence*
- *International Journal of Computational Intelligence Systems*
- *International Journal of Computational Intelligence and Applications*
- *International Journal of Computational Intelligence and Organizations*
- *International Journal of Computational Intelligence Research*
- *International Journal of Computational Intelligence Theory and Practice*

## References

1. G. Casalino, Computational intelligence for smart laser materials processing. Opt. Laser Technol. **100**, 165–175 (2018)
2. M.N.O. Sadiku, S.M. Musa, O.M. Musa, Computational intelligence: a primer. Int. J. Adv. Eng. Technol. **3**(4), 34–36 (2019)
3. M.N.O. Sadiku, J. Foreman, S.M. Musa, Computational intelligence. Eur. Sci. J. **14**(21), 56–60 (2018)
4. What is computational intelligence? https://cis.ieee.org/about/what-is-ci
5. D. Hecht, Applications of machine learning and computational intelligence to drug discovery and development. Drug Dev. Res. **72**, 53–65 (2011)
6. U. Kumari, Soft computing applications: a perspective view, in *Proceedings of the 2nd International Conference on Communication and Electronics Systems* (2017), pp. 787–789
7. M.N.O. Sadiku, Y. Wang, S. Cui, S.M. Musa, Soft computing: an introduction. Int. J. Adv. Res. Comp. Sci. Soft. Eng. **8**(6), 63–65 (2018)
8. I. Fister et al., Computational intelligence in sports: challenges and opportunities within a new research domain. Appl. Math. Comput. **262**, 178–186 (2015)
9. R.V. Kulkarni, A. Forster, G.K. Venayagamoorthy, Computational intelligence in wireless sensor networks: a survey. IEEE Commun. Surv. Tutor. **13**(1), 68–96 (2011)
10. Y.F. Hernández-Julio et al., Framework for the development of business intelligence using computational intelligence and service-oriented architecture, in *Proceedings of the 12th Iberian Conference on Information Systems and Technologies (CISTI)* (2017), pp. 1–7
11. P. Isasi et al., Applied computational intelligence for finance and economics. Comput. Intell. **23**(2), 111–116 (2007)
12. A.E. Ruano, Computational intelligence in control, in *Proceedings of the 19th World Congress The International Federation of Automatic Control*, Cape Town, South Africa (2014), pp. 8867–8878
13. G. Cosma et al., A survey on computational intelligence approaches for predictive modeling in prostate cancer. Expert Syst. Appl. **70**, 1–19 (2017)
14. M.A. Jabbar, B.L. Deekshatulu, P. Chandra, Computational intelligence technique for early diagnosis of heart disease, in *Proceedings of IEEE International Conference on Engineering and Technology (ICETECH)*, Coimbatore, India (2015)

15. S. Wiwanitkit, V. Wiwanitkit, Computational intelligence in tropical medicine. Asian Pac. J. Trop. Biomed. **6**(4), 350–252 (2016)
16. C.H. Liu, C.K. Ting, Computational intelligence in music composition: a survey. IEEE Trans. Emerg. Top. Comput. Intell. **1**(1), 2–15 (2017)
17. D. Apostolou et al., Facilitating creativity in collaborative work with computational intelligence software. IEEE Comput. Intell. Mag., 29–40 (2016)
18. R. Iqbala et al., Big data analytics: computational intelligence techniques and application areas. Technol. Forecast. Soc. Chang. (2018)
19. M. Li, Application of computational intelligence in modeling and optimization of HVAC systems, Masters Thesis, The University of Iowa (2009)
20. T. Hanne, R. Dornberger, *Computational Intelligence in Logistics and Supply Chain Management* (Springer, 2017)
21. E. Diamant, Computational intelligence: are you crazy? Since when has intelligence become computational? in *Proceedings of IEEE Symposium Series on Computational Intelligence (SSCI)* (2016)
22. A. Kalantari et al., Computational intelligence approaches for classification of medical data: state-of-the-art, future challenges and research directions. Neurocomputing, 1–21 (2017)
23. M. Zhang, Experience of teaching computational intelligence in an undergraduate level course. IEEE Comput. Intell. Mag., 57–59 (2011)
24. A. Konar, *Computational Intelligence: Principles, Techniques and Applications* (Springer, 2005)
25. W. Pedrycz, *Computational Intelligence: An Introduction* (CRC Press, Boca Raton, FL, 1998)
26. D. Poole, A. Mackworth, R. Goebel, *Computational Intelligence – A Logical Approach* (Oxford University Press, New York, 1998)
27. Z. Chen, *Computational Intelligence for Decision Support* (CRC Press, Boca Raton, 2000)
28. A. Kusiak, *Computational Intelligence in Design and Manufacturing* (Wiley-Interscience, 2000)
29. D. Andina, D.T. Pham, *Computational Intelligence: For Engineering and Manufacturing* (Springer, 2007)
30. A.T. Azar, S. Vaidyanathan, *Computational Intelligence Applications in Modeling and Control* (Springer, 2015)
31. R. Kruse et al., *Computational Intelligence* (Springer, 2012)
32. W. Pedrycz, S.M. Chen, *Information Granularity, Big Data, and Computational Intelligence* (Springer, 2014)
33. A. Kordon, *Applying Computational Intelligence: How to Create Value* (Springer, 2010)
34. N. Ansari, E. Hou, *Computational Intelligence for Optimization* (Kluwer Academic Publishers, London, 1997)
35. R. Srivastava et al. (eds.), *Computational Intelligence for Machine Learning and Healthcare Informatics* (De Gruyter, 2020)
36. H. Deng et al. (eds.), *Artificial Intelligence and Computational Intelligence* (Springer, 2012)
37. M. Gui et al., *Computational Intelligence and Soft Computing Applications in Healthcare Management Science* (IGI Global, 2020)
38. C. Piñeiro et al. (eds.), *Trends in Mathematics and Computational Intelligence* (Springer, 2019)
39. A. T. Azar, S. Vaidyanathan (eds.), *Computational Intelligence Applications in Modeling and Control* (Springer, 2015)
40. R.C. Eberhart, Y. Shi, *Computational Intelligence: Concepts to Implementations* (Elsevier, 2012)
41. R. Kruse et al., *Computational Intelligence: A Methodological Introduction*, 2nd edn. (Springer, 2013)
42. K. Suzuki (ed.), *Computational Intelligence in Biomedical Imaging* (Springer, 2014)
43. P.P. Angelov, *Handbook on Computational Intelligence (in 2 Volumes)* (World Scientific, 2016)
44. J. Kacprzyk, W. Pedrycz, *Springer Handbook on Computational Intelligence* (Springer, 2015)
45. A.P. Engelbrecht, *Computational Intelligence: An Introduction*, 2nd edn. (Wiley, 2007)
46. A. Chatterjee, P. Siarry, *Computational Intelligence in Image Processing* (Springer, 2013)
47. J. Fulcher, L. C. Jain (eds.), *Computational Intelligence: A Compendium* (Springer, 2008)

48. K. Pancerz, E. Zaitseva (eds.), *Computational Intelligence, Medicine and Biology* (Springer, 2015)
49. H. K. Lam, S. S. H. Ling, H. T. Nguyen (eds.), *Computational Intelligence and its Applications: Evolutionary Computation, Fuzzy Logic, Neural Network and Support Vector Machine Techniques* (Imperial College Press, London, 2012)
50. T. Islam et al. (eds.), *Computational Intelligence Techniques and Applications* (Springer, 2014)
51. L. Jain, P. De Wilde (eds.), *Practical Applications of Computational Intelligence Techniques* (Springer, 2001)
52. M. G. Negoita, B. Reusch (eds.), *Real World Applications of Computational Intelligence* (Springer, 2005)
53. V. Palade, C.D. Bocaniala, *Computational Intelligence in Fault Diagnosis* (Springer, 2000)
54. L.N. Castro, L.N. De Castro, J. Timmis, *Artificial Immune Systems: A New Computational Intelligence Approach* (Springer, 2002)
55. L. Wang, X. Fu, *Data Mining with Computational Intelligence* (Springer, 2005)

# Chapter 13
# Digital Intelligence

**Contents**



Embracing digital is inevitable as that is now part of the business venture.

—Pearl Zhu

## 13.1 Introduction

Digital is the language of today. Digital technologies are changing the way we think and live. They are raping the world around us. They have a firm grasp on practically every aspect of human life. We interact with digital technologies when we work, play, learn, communicate, and do business. To survive in the digital era, digital skills are more important than ever before. Digital technologies give us superhuman abilities and we need to know how to use these abilities wisely.

The exposure to digital technologies is not limited to adults. Our children are the most vulnerable. Children spend several hours a day in front of a screen. They can gain access to the online world from the palm of their hand. By the age of nine, some children already have their own mobile device. Children know how to navigate their way on the technological highway like no other. Figure 13.1 illustrates the impact of technology on children [1]. Children with a high DQ have the potential for becoming critical thinkers. It seems expedient then that children need to understand how to behave in the online world. Parents should understand their



M. N. O. Sadiku, S. M. Musa, *A Primer on Multiple Intelligences*,
https://doi.org/10.1007/978-3-030-77584-1_13

**Fig. 13.1** Digital intelligence can protect children from online threats [1]

responsibility when it comes to ensuring digital intelligence. Also, it is imperative that we empower children to become good digital citizens who are responsible users of technology.

Our traditional approach to learning has proved completely inadequate. As the workplace becomes increasingly dominated by digital technology, we need new ways of thinking and working to be productive and effective. We need a new approach focused on aligning our digital tools. We need a new type of intelligence to help us navigate the challenges and opportunities of digital world [2].

Digital intelligence refers to the combination of social, emotional, and cognitive abilities that enable individuals to adapt to the demands of life in the digital age. It is the use of generated data by digital technologies to improve the user experience. This is the kind of intelligence everybody ought to have in the twenty-first century.

This chapter is an introduction on digital intelligence. It begins by discussing what digital intelligence is all about. It describes the digital citizenship competencies. It presents some applications of digital intelligence. It highlights the benefits and challenges of digital intelligence. It covers global digital intelligence and how digital intelligence can be measured. It concludes with some comments.

## 13.2 Concept of Digital Intelligence

Intelligence is often regarded as the ability to solve problems. A newly emerging form of intelligence is digital intelligence. Digital intelligence is a pressing concept at the intersection of technology and society. In the hyper-connected world, digital intelligence is an important skill for future generations.

Digital intelligence (also known as digital intelligence quotient (DQ)) refers to an all-encompassing set of technical, cognitive, social, and emotional competencies that enable individuals to face the challenges of digital life. It belongs to meta-intelligence—one that is composed of many constituent intelligences. Digital intelligence addresses the what, why, where, when, who, how, and how much of digital technology to improve our operational efficiency and performance. It is about putting the data to use immediately. It is the intersection of big data with big science.

The digital intelligence (DQ) is similar to the intelligence quotient (IQ), emotional quotient (EQ), and social quotient (SQ). Unlike IQ, which is commonly viewed as a genetically determined intelligence, DQ is something that needs to be developed, especially from childhood. Digital intelligence basically deals with our relationship with technology, while emotional intelligence is about our relationship with others. DQ is the combination of skills, abilities, and knowledge that are essential to digital living.

The idea of "digital intelligence" was first introduced in 2016 by Yuhyun Park in his published work in the World Economic Forum. Since then, the DQ framework has been widely used by local and international organizations. DQ may be regarded as part of the globalization of markets, the digitization of work, and development of the Fourth Industrial Revolution. Some established companies have embraced digital intelligence as core to their overall corporate strategies. Examples of such companies include Google, Amazon, eBay, Polaris Industries, and Telegraph Media Group.

Digital intelligence helps individuals to acquire and apply new knowledge and skills related to digital technologies: social, mobile, analytics, artificial intelligence, cloud, blockchain, and cybersecurity. We develop our digital intelligence as we interact with these different digital technologies [3].

## 13.3 Digital Intelligence Competencies

The concept of digital intelligence encompasses eight interconnected areas: digital identity, digital use, digital safety, digital security, digital emotional intelligence, digital communication, digital literacy, and digital rights. These abilities are also known DQ digital citizenship competencies and are explained as follows [4].

- *Digital Identity:* The ability to create and manage one's online identity and reputation (the so-called one's online self). This includes awareness of one's online identity and control of short- and long-term impact of one's online identity.
- *Digital Use:* The ability to use digital devices and media, including the mastery of control in order to achieve a healthy balance between life online and offline.
- *Digital Safety*: The ability to manage risks online (e.g., cyberbullying, grooming, radicalization) as well as problematic content (e.g., violence and obscenity), and to avoid and limit these risks.
- *Digital Security*: The ability to detect cyber threats (e.g., hacking, scams, malware), to understand best practices, and to use suitable security tools for data protection.

- *Digital Emotional Intelligence*: The ability to be empathetic and build good relationships with others online.
- *Digital Communication:* The ability to communicate and collaborate with others using digital technologies and media.
- *Digital Literacy:* The ability to find, evaluate, utilize, share, and create content as well as competency in computational thinking.
- *Digital Rights:* The ability to understand and uphold personal and legal rights.

These eight key skills have been identified as making up a person's digital IQ. They are illustrated in Fig. 13.2. One may also add more competencies [5].

- *Screen Time Management:* The ability to manage one's screen time, multitasking, and participation in online games and social media with self-control.
- *Cyberbullying Management*: The ability to detect situations of cyberbullying and handle them wisely.
- *Cybersecurity Management:* The ability to protect one's data by creating strong passwords and to manage various cyberattacks.
- *Privacy Management:* The ability to discretely handle all personal information shared online in order to protect one's and others' privacy.
- *Critical Thinking:* The ability to distinguish between true and false information, good and harmful content, and trustworthy and questionable contacts online.
- *Digital Footprints*: The ability to understand the nature of digital footprints and their real-life consequences and to manage them responsibly.

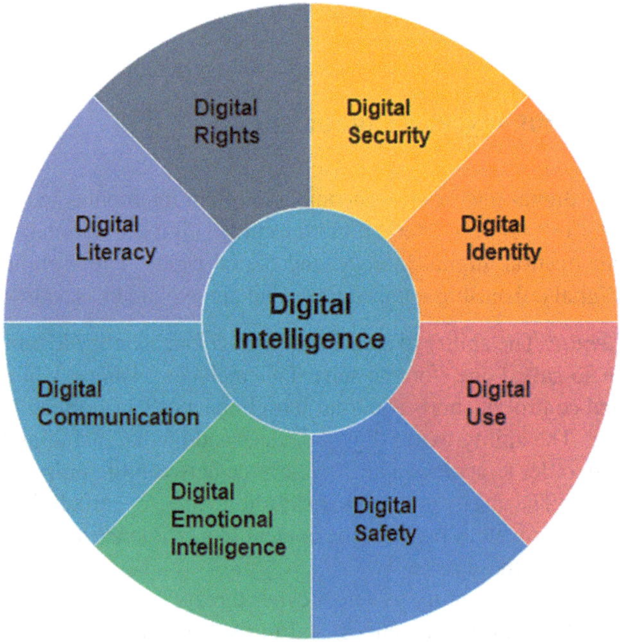


**Fig. 13.2** The eight competencies of digital intelligence

- *Digital Empathy:* The ability to show empathy towards one's own and others' needs and feelings online.

It has been recommended that everybody, particularly the young learners, needs to be armed with these competencies, rooted in universal moral values, in order to be successful in the digital world and become good digital citizens [6].

## 13.4 Applications

Digital technologies are rapidly changing the world. Digital intelligence is increasingly relevant to many areas such as school education, business, marketing, law enforcement, and culture.

- *Education:* Digital technologies support the process of learning. Digital intelligence may be regarded as a new concept in context of future school of education. The current school system has been undergoing changes at the global level due to the pressure of digital technologies. For educators who prepare future generations for engagement in this rapidly evolving world environment, it is critical they embrace the emerging digital technologies [1, 7]. Everyone around the world, starting with our children, must have equal opportunity to thrive in this digital age. Parents also must assume some responsibility to keep their children safe online. A parent can help their child with digital intelligence in the following ways [8]: (1) Help them think through, (2) Engage in monitoring, (3) Talk about experiences, (4) Limit screen time, (5) Help them in future decisions.
- *Business:* In business, the goal of digital transformation is to create a smarter, faster business that can better anticipate and meet its customers' needs. Transforming a business to become fully digital is not easy. For a business to survive, it must go from being insights-influenced to being insights-driven. Focusing on increasing digital intelligence in specific areas will increase the chances that the digital transformation will succeed. In business, digital intelligence refers to the information obtained through technologies and making use of them as an online marketing strategy. It involves understanding your customers and how they're using your website. As shown in Fig. 13.3, digital intelligence is one of seven enablers that can help financial service providers renew the customer experience in the digital era. Digital intelligence is concerned with driving the right business outcomes, especially improved customer experience, increased revenues, faster decision-making, greater profit, reduced cost, and improved brand preference and trust. It involves understanding your customers and how they are using your website. It empowers companies to connect to more customers. DQ can transform digital data into actionable, customer-centric insights. Several industries have embraced digital intelligence including banking, retail, and healthcare.
- *Marketing:* Digital intelligence is the backbone of customer experience management. The rapid technological innovations have transformed the structure of

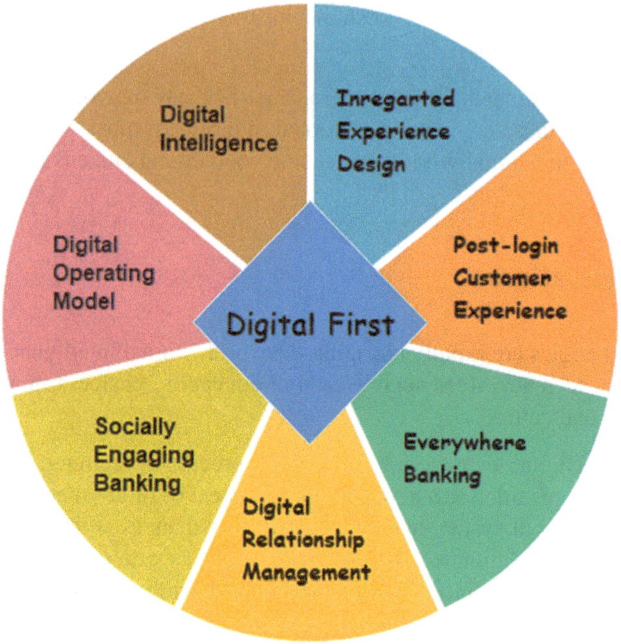


**Fig. 13.3** Digital intelligence is one of seven enablers that can help financial service providers renew the customer experience

markets. Therefore, all digital marketers should understand what digital intelligence is all about and how it can be leveraged to deliver a more effective customer experience. This can help get the right message at the right time, to the right audience through the right channel. Digital marketers across all industries can achieve success if they focus on the following 5 Cs of digital marketing: consumer/customer/client, context, content, commerce, and convergence [9].

- *Human Resources:* The HR department attracts talents, defines new management rules, and promotes corporate values. It holds the key to a successful digital transformation. This requires that the HR department work together closely with the other departments. The role of HR department is more essential than ever [10].
- *Library:* Libraries are playing an increasingly important role in smart cities. Libraries serve as indispensable partners for cities to roll out initiatives and connect residents with crucial services. Cities use libraries as centers to share information about initiatives to the public.
- *Forensic Casework:* There has been an increased use of technology as a medium for engaging in illegal activities. Companies are just realizing the tip of the iceberg when it comes to computer forensics. Increasing operational efficiencies is one of the biggest challenges facing forensics today.
- *Law Enforcement:* Crimes are becoming increasingly complex and influenced by technology. Cybercrime is the most rapidly growing form of crime, using cyberspace to conduct classic criminal activity at scale: stealing money, organizing

narcotics, people smuggling, and accessing credit cards and other personal details of potential victims. Digital intelligence solutions enable legal professionals to easily digitize, edit, enhance, and manage their legal documents. This creates a positive customer experience that improves satisfaction, increases revenues, and strengthens engagement.

- *Digital Cultural Intelligence* (Digital CQ): This is at the backbone of every strategic decision we make. Digital CQ is essentially what we say and how well we say it. It refers to an organization's ability to clearly articulate the brand promise.

Other applications of digital intelligence include information warfare, digital evidence, digital economy, world of disabled persons, and oil and gas.

## 13.5 Benefits

Digital intelligence has been identified as an effective means of improving digital citizenship by the World Economic Forum. It has been recognized as a new democratic license to operate. It encompasses the skills needed to use technology more effectively. Almost all companies are aware of the necessity to engage in digital transformation. Organizational ability to mobilize these digital technologies wisely leads to digital creativity.

We all can benefit from an understanding and improvement of our DQ since it helps us to get ahead of technology which helps us to get ahead in life. Digital intelligence enables individuals to adapt to the demands of digital life. It is key to a company's survival. Thus, digital intelligence is a key competence for the future of work for both employers and employees. Recruiting employees with high digital intelligence is a near guarantee of success today for an organization. Digital intelligence also bridges the gap between expectations and business outcomes such as improved customer experience, increased revenues, and reduced cost to serve.

Helping children develop digital intelligence is vitally important to prepare them to face the challenges, risks, and demands of the digital world. The preparation will also provide them future skills needed for life in a digitized and hyperconnected world.

## 13.6 Challenges

Digital intelligence requires that every individual faces the challenges and adapts to the demands of life in the digital world. Today, keeping up with the rapidly changing technological environment is a major challenge. Since managing one's digital intelligence is difficult for adults, most of the DQ initiatives available are about focused on youths. Transitioning to digital world puts some pressures on youths. These include reduced digital empathy, excessive screen time, digital addiction, cyberbullying, online grooming, digital identity theft, online privacy mismanagement, exposure to digital misinformation operations, etc. [11].

There are no globally shared understanding of what terms like digital skills, digital literacy, digital readiness, and digital intelligence mean. Some critics argue that digital intelligence is not really a form of intelligence. The distinction between an intelligence and a skill is often a source of confusion. Digital transformation, also known as digitalization, puts a mounting pressure on organizations and businesses to adapt, change, and innovate in order to remain competitive [12]. Since there are different digital technologies for different purposes, DQ places heavy demands on individual workers. Modern workers must be able to integrate the various workplace needs of speed, mobility, multitasking, and efficiency by working ubiquitously—anywhere, anytime, and through one device.

Education systems around the world are typically not well equipped to integrate digital intelligence in their school curricula. Parents are concerned about how their children might deal with problems often encountered online. Educators, law enforcers, government, the media, and parents should understand the implications of a child gaining access to the online world from the palm of his or her hand. The consequences can be devastating and can cause adolescent suicide. These challenges should be addressed by the society in a collective manner.

## 13.7 Global Digital Intelligence

Digital intelligence is grounded in universal moral values that enable individuals to face the challenges of the digital world. Digital transformation is a key component of the national agenda for many nations. Several nations have pledged their support for developing digital intelligence including the United States, United Kingdom, Argentina, India, Australia, Singapore, Korea, and Taiwan. A Coalition for Digital Intelligence (CDI) was launched in 2018 and sponsored by the World Economic Forum (WEF). Member nations of WEF include the USA, Argentina, India, Australia, Singapore, Korea, and Taiwan, while member companies include the Varkey Foundation, Google, and LG. The CDI is an international network of organizations that aims to improve digital intelligence by coordinating the implementation of the DQ framework, to make a greater impact on the advancement of digital skills. It ensures that both technology and education sectors work together and the world moves towards universal digital intelligence. Figure 13.4 depicts the mission of CDI.

Setting standards for global digital intelligence is to ensure that everyone around the world has equal opportunity to thrive and survive in this digital age. These standards are fundamental for future readiness. One of such standards is for digital literacy, skills, and readiness. Billions of dollars are invested every year in improving digital literacy and skills.

Each nation should develop a master plan for building a digital education curriculum.

In the near future, schools, from elementary to higher education, will be responsible for helping students develop their digital intelligence. Such preparation will help with potential technological threats such as digital addiction, cyberbullying, radicalization, online grooming, online sexual behaviors, digital identity theft, online privacy mismanagement, meeting online strangers, and exposure to digital disinformation.

The DQ institute and similar companies have developed teaching materials to assist educators in introducing DQ skills to students worldwide. For example, in South Korea, UNICEF is supporting the introduction of DQ in schools to help children manage their screen time, boost their critical thinking, and use the Internet wisely and safely.

DQ Every Child is a global movement that seeks to empower every child in the world with DQ digital citizenship. It makes use of an online gamified platform to educate and assess children with digital citizenship skills. The movement was created in Singapore and has reached several nations such as Korea, Nigeria, Turkey, Mexico, and India.

By 2019, the movement has reached more than 700,000 children in 107 countries and the initiative has been translated into 21 languages [1, 6]. Digital Citizenship is another global movement for the young generation, the older generation, corporate professionals, parents, and educators.

There is also a real need for the widest possible international mutual legal recognition of certain clear classes of criminal offence that are cyber enabled.

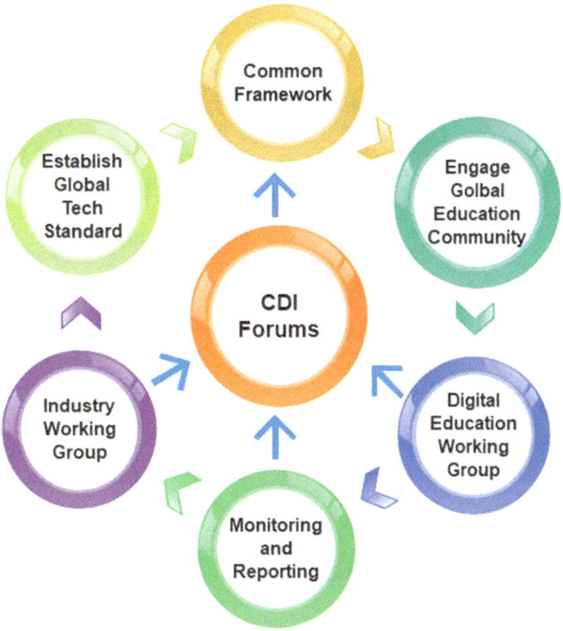


**Fig. 13.4** The mission of the Coalition for Digital Intelligence (CDI)

## 13.8 Assessing Digital Intelligence

For more than a century, psychologists have used certain metrics to evaluate abilities and ultimately arrive at a human intelligence quotient (IQ). Unlike IQ, which is commonly viewed as a genetically determined intelligence, DQ is something that needs to be built. DQ is measurable and highly learnable. DQ score shows how well individuals can use digital technology and media in responsible and effective ways. Figure 13.5 shows how DQ evaluates four outcomes.

Take the following simple digital test and find out just how savvy you are [13]. The test allows you to self-assess your digital quotient.

1. What is your tolerance for risk?
   (a) Low
   (b) Medium
   (c) High
2. Do you have a bold business vision?
   (a) For Sure
   (b) No
   (c) Somewhat
3. Are you open to listen to new ideas from other business units?
   (a) No
   (b) Absolutely
   (c) Sometimes

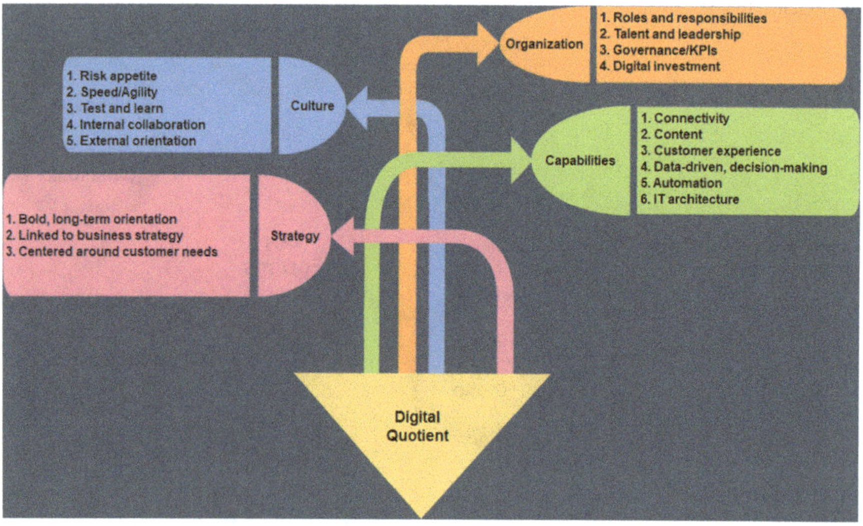


**Fig. 13.5** Digital quotient evaluates four outcomes

4. Do you believe digital is the sole remit of marketing?
   (a) No way
   (b) Probably
   (c) Yes
5. Will your business adopt new digital processes?
   (a) For Sure
   (b) Somewhat
   (c) No
6. If you plan on adopting new digital technologies, what would these comprise?
   (a) Marketing Automation
   (b) ERP/CRM
   (c) Cloud
7. Are you aware of ransomware and the impact it could have on your business?
   (a) Somewhat
   (b) No
   (c) For Sure
8. How cloud savvy is your organization?
   (a) Medium
   (b) High
   (c) Low

A more advanced digital intelligence quotient (DIQ) scale questionnaire encompasses the digital identity, digital use, digital safety, digital security, digital emotional intelligence, digital communication, digital literacy, and digital rights [4].

## 13.9 Conclusion

We live in the digital age, where digital technologies have transformed the way organizations and businesses operate. Since organizations cannot cope with the volume of information available, they need the ability to accelerate the process to get to actionable intelligence. Digital intelligence is the practice of capturing, managing, and analyzing customer data and insights to deliver a holistic view of customers' digital interactions.

Digital intelligence practitioners use data and analytics to delight customers with optimal experiences wherever they digitally interact with their firms.

Digital intelligence follows other such phenomena as the Internet of things, big data, artificial intelligence, blockchain, and cloud computing, which are becoming part of our daily lives. It is here to stay and there is truly nowhere to hide. As our lives become increasingly hyper-connected, societies the world over will depend on digital intelligence. Digital intelligence will be what matters in the future.

To help prepare students for the workplace, academic institutions should add to their curricula courses that embed digital intelligence. For more information about digital intelligence, one should consult the books in [14–19].

## References


1. Forget IQ. Digital intelligence will be what matters in the future, https://www.weforum.org/agenda/2018/02/digital-intelligence-internet-safety-future/
2. M.N.O. Sadiku, S.M. Musa, A. Ajayi-Majebi, Digital intelligence: a primer. Int. J. Trend Res. Develop. **7**(4) (2020)
3. I. Boughzala, Digital intelligence: a key competence for the future of work (2019), https://www.aacsb.edu/blog/2019/march/digital-intelligence-a-key-competence-for-the-future-of-work#gsc.tab=0
4. K. Na-Nan, T. Roopleam, N. Wongsuwan, Validation of a digital intelligence quotient questionnaire for employee of small and medium-sized Thai enterprises using exploratory and confirmatory factor analysis. Kybernetes **49**(5), 1465–1483 (2019)
5. DQ Institute, Digital intelligence (DQ): A conceptual framework and methodology for teaching and measuring digital citizenship (2017), https://www.dqinstitute.org/wp-content/uploads/2017/08/DQ-Framework-White-Paper-Ver1-31Aug17.pdf
6. Digital intelligence, Wikipedia, the free encyclopedia, https://en.wikipedia.org/wiki/Digital_intelligence
7. J. Dostál et al., Digital Intelligence - New concept in context of future school of education, in *Proceedings of ICERI2017 Conference*, Seville, Spain (2017)
8. Y. Williams, Why digital intelligence is important and 5 ways parents can help (2017), https://www.huffpost.com/entry/why-digital-intelligence-is-important-and-5-ways-parents_b_5a315fd9e4b04bd8793e9671
9. C. Iliuţă, I. Matei, Digital intelligence and digital marketing effectiveness. Holist. Market. Manag. J. **7**(1), 12–17 (2017)
10. B. Protais, What is your digital quotient? (2019), https://en.blog.businessdecision.com/what-is-your-digital-quotient/
11. Q. Terry, One in three kids use an iPad before they can speak. This is the emergence of Digital Intelligence (2018), https://www.linkedin.com/pulse/one-three-kids-use-ipad-before-can-speak-emergence-digital-terry
12. I. Boughzala, M. Garmaki, O. Chourabi, Understanding how digital intelligence contributes to digital creativity and digital transformation: a systematic literature review, in *Proceedings of the 53rd Hawaii International Conference on System Sciences* (2020), pp. 320–329
13. Raise your digital quotient, https://www.mckinsey.com/featured-insights/digital-disruption/raise-your-digital-quotient
14. S.L. Mithas, *Digital Intelligence: What Every Smart Manager Must Have for Success in an Information Age*, 3rd edn. (Finerplanet, 2016)
15. M. Johnson, *Cyber Crime, Security and Digital Intelligence* (Routledge, London, 2013)
16. M. Yildiz, *Digital Intelligence: A Framework to Digital Transformation Capabilities* (Independently Published, 2019)
17. Digital Intelligence a Complete Guide - 2020 Edition. Emereo Pty Limited, 2020
18. W.G. Perry, *Information Warfare: Assuring Digital Intelligence Collection* (Amazon Digital Services, 2019)
19. I.P. Australia, *Puts Digital Intelligence to Work* (World Intellectual Property Organization, 2018)

# Part III
# Other Intelligences

# Chapter 14
# Business Intelligence

## Contents



> Business Intelligence is the process of utilizing organizational data, technology, analytics, and the knowledge of subject matter experts to create data-driven decisions via dashboards, reports, alerts, and ad-hoc analysis.
>
> —Anonymous

## 14.1 Introduction

Businesses constitute the pillar of American economy. They play a crucial role in creating jobs, increasing production, and contributing to the economy in many ways. To turn any business into a profit-making enterprise requires that the work force make wise decisions which depend on the available information. This is where business intelligence comes in.

The concept of business intelligence (BI) refers to the tools and systems that play a major role in the planning process of an organization. The tools represent BI in customer support, customer profiling, market research, product profitability, and business analysis. They are regarded as technology that enhances efficiency of business operations. The intelligent systems help in decision-making process by gathering, storing, and analyzing data [1]. They help business executives, managers, and others make more informed, better, and faster decisions. The objective of BI is to



M. N. O. Sadiku, S. M. Musa, *A Primer on Multiple Intelligences*,
https://doi.org/10.1007/978-3-030-77584-1_14

improve the timeliness and quality of data for planners and decision-makers. BI technologies are used for acquiring and transforming raw data into meaningful information for the purpose of business analysis. Their common functions include real-time reporting, processing, data mining, and business performance management [2]. BI may be regarded as the application of human intelligence in business activities.

While BI means different things to different people, it may be seen as an umbrella term that finds its roots in areas such as decision-making, information system, and data warehousing. Many large- and medium-sized companies now use BI.

This chapter provides how companies are using business intelligence to improve the effectiveness of decision-making. It begins by defining business intelligence. It describes various components of business intelligence. It discusses business intelligence analytics and tools. It presents some applications of business intelligence. It highlights some benefits and challenges of business intelligence. The last section concludes with comments.

## 14.2 What Is Business Intelligence?

The concept of "business intelligence" (BI) was first used by Howard Dresner in 1988. A short history of business intelligence is shown in Fig. 14.1. The term BI refers to the tools and systems that play a major role in the planning process of an organization. The tools represent BI in customer support, customer profiling, market research, product profitability, and business analysis. The main objective of business intelligence is to enable business managers to have easy assess to data, be able to conduct analysis, allow them to convert data into useful knowledge, and then make faster, better decisions [3]. BI software supports reporting, interactive analyses, visualization, and statistical data mining. People use business intelligence for many reasons such as the need to seek improvements in decision-making and competition and the need to seek efficiency.

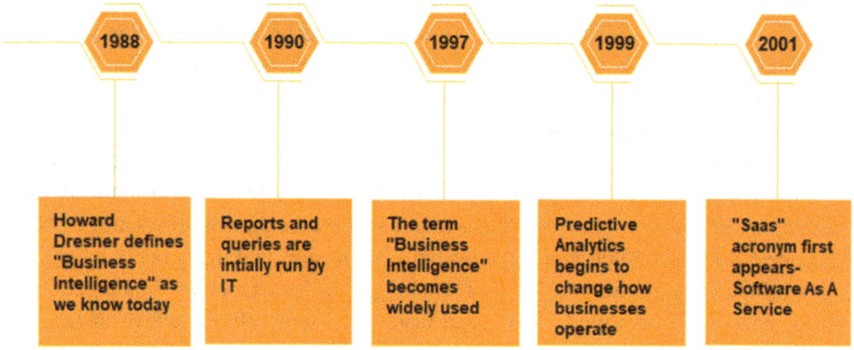


**Fig. 14.1** A short history of business intelligence

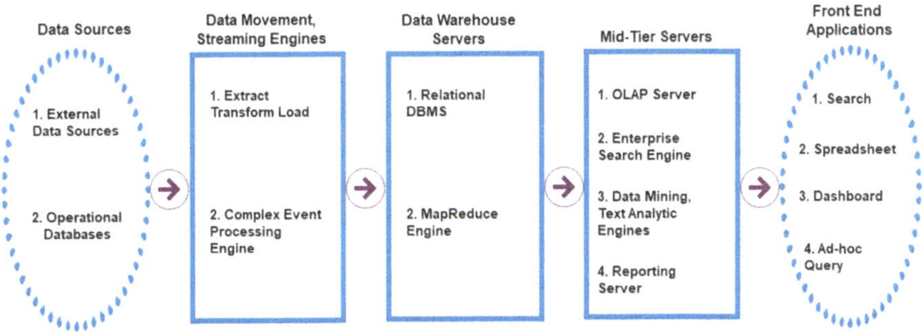


**Fig. 14.2** A typical business intelligence architecture

Business intelligence (BI) refers to technologies, applications, and practices for the collection, integration, analysis, and presentation of business information. It is a broad term to describe a set of methods, processes, and technologies that transform raw data into meaningful and useful information to support business operation. The purpose of BI is to support better business decision-making [4]. Business intelligence also refers to a broad category of analytics, data warehousing and visualization tools. The purpose of BI is to support better business decision-making. Although BI is designed for the business environment, it can be used by government, universities, healthcare, and security. Figure 14.2 shows a typical business intelligence architecture.

## 14.3 Components of BI

BI system is an integrated set of tools, technologies, and programmed products, which are used to collect, integrate, analyze, and share data. BI system has following components [5].

- *Data Mining*: Sorting through large data sets using databases, statistics, and machine learning to identify trends and establish relationships.
- *Big Data*: We live in era of big data. BI applications extract, organize, analyze, and report large volumes of data to aid strategic and managerial decision-making.
- *Querying:* A request for specific data or information from a database.
- *Data Preparation:* The process of combining and structuring data in order to prepare it for analysis.
- *Reporting:* Sharing operating and financial data analysis with decision-makers so they can draw conclusions and make decisions.
- *Benchmarking:* Comparing current business processes and performance metrics to historical data to track performance against industry bests.
- *Descriptive Analytics*: The interpretation of historical data to draw comparisons and better understand changes that have occurred in a business.

- *Statistical Analysis:* Collecting the results from descriptive analytics and applying statistics in order to identify trends.
- *On-line Analytical Processing* (OLAP): These are BI tools that are used to analyze large volumes of historical data with drill-down functionality. Information is stored in OLAP cubes and provides a multidimensional view of data.
- *Data Visualization:* This is the use of visual representation to explore, analyze, and interpret large amounts of data. It incorporates graphs, charts, Geographic Information Systems (GIS), Graphical User Interface (GUI), virtual reality, video, and animation. Data visualization technologies transform large quantitative data sets into useful information so that individuals can understand.
- *Data Warehouse*: This provides storing of aggregated and analyzed data. It allows reports to be produced at a reasonable cost for community.
- *Data Management and Integration*: This component prepares data to be ready for analysis and reports. It involves analyzing, reporting, and presenting tools such as on-line analytical processing (OLAP) tools which allow users access and which analyze and model business problems and share information.
- *Mobile BI*: Software that optimizes desktop business intelligence for mobile devices. Mobile BI solutions are available with voice-enabled access.
- *Collaborative BI:* This is the merging of business intelligence software with collaboration tools in order to streamline the sharing process.

Some of these components are shown in Fig. 14.3.

Business Intelligence (BI) is a fast growing field with significant benefits and opportunities for healthcare workers. The healthcare workers need fast, responsive, and user-friendly BI software to maximize the value of their data and support critical areas of decision-making. BI tools are must-haves for any long-term and sustainable analytics foundation.

The seven steps to developing BI systems are [6]:

1. Make sure your data is clean.
2. Train users effectively.
3. Deploy quickly, then adjust as you go. Deliver reports that provide the most value quickly, and then tweak them.
4. Take an integrated approach to building your data warehouse from the beginning. Make sure you are not locking yourself into an unworkable data strategy further down the road.
5. Define return on investment (ROI) clearly before you start. Outline the specific benefits you expect to achieve, then do a reality check every quarter or 6 months.
6. Focus on business objectives.
7. Do not buy business intelligence software because you think you need it. Deploy BI with the idea that there are numbers out there that you need to find, and know roughly where they might be.

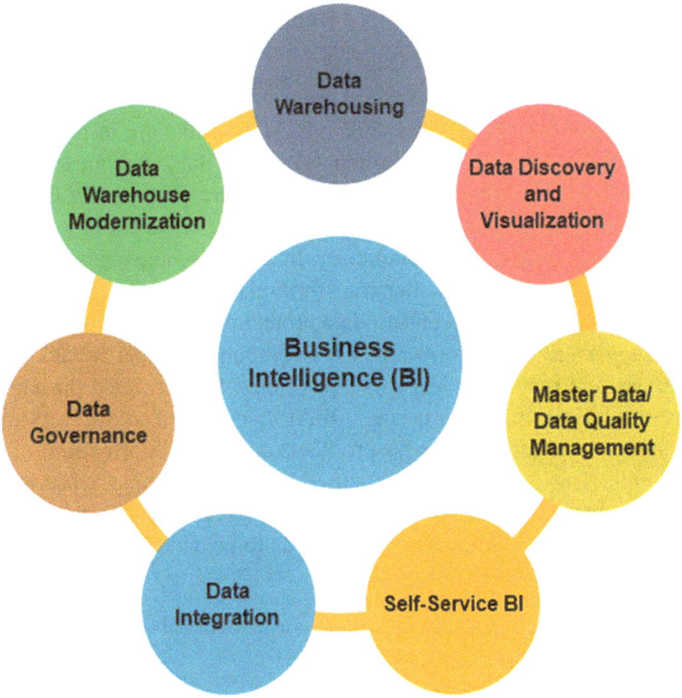


**Fig. 14.3** The components of business intelligence

## 14.4 Business Intelligence Analytics

Business intelligence is increasingly becoming necessary for business leaders to visualize, analyze, and prepare their strategic planning. In recent years, several organizations have implemented business intelligence and analytics for big data solutions. Busines intelligence and business analytics go hand in hand. While business intelligence focuses mainly on what is happening in your organization, business analytics includes solutions that help you leverage that insight to plan for the future. Business data analysis turns raw information into actionable insights, helping one to make a better decision.

Different types of data analytics can help organizations better serve patients in several ways [7]:

- *Descriptive Analytics*: Shows what is happening or has been happening.
- *Diagnostic Analytics*: Points to relationships and causes like tracing heart disease back to poor diets.
- *Predictive Analytics*: Indicates what will likely happen.
- *Prescriptive Analytics*: Recommends specific actions in response to individual patient symptoms.

BI system is an integrated set of tools, technologies, and programmed products, which are used to collect, integrate, analyze, and share data. BI system has following components [5].

- *Extraction-Transformation-Load* (ETL): ETL tools that are responsible for data transfer from operational or transaction systems to data warehouses.
- *Data Warehouse*: Data Warehouse (DW) is the core of any solid BI solution. It provides storing of aggregated and analyzed data. It allows reports to be produced at a reasonable cost for community. In data warehousing, you create stores of informational data which is then transformed for end-user decision-making. A good data warehouse should ensure data protection and security.
- *Data Management and Integration:* This component prepares data to be ready for analysis and reports. It involves analyzing, reporting, and presenting tools such as on-line analytical processing (OLAP) tools which allow users access and which analyze and model business problems and share information.
- *Presentation:* This component is a focused layer which brings sense-making and decision support directly to users. Presentation layers include customized graphical and multimedia interfaces or dashboards to provide users with information in a comfortable and accessible form.

Business-specific applications require all components above to work together to realize a complete solution.

## 14.5 BI Tools

Business intelligence tools enable leaders to manage data, predict the future of technology, and make good decisions. They perform data analysis and create reports, summaries, graphs, and charts to provide users with detailed information on the nature of the business. There are plenty of them in the market [8].

- *MS Excel*: The most commonly used tool for BI is MS Excel because it is commonly available and can easily be used to display data.
- *JasperReports*: This is an open-source BI tool that was developed by Jaspersoft. It has learning sources including training and webinars.
- *Pentaho*: This is a suite of tools for data analysis and data mining. It provides a platform for cloud-based visualization of data.
- *SpagoBI*: It is part of the SpagoWorld open-source initiative. It consists of 15 components which include reporting, OLAP, dashboards, and data mining.
- *Palo/Jedox*: It is designed for a business suite for planning and reporting. It was developed by German company Jedox. Add-ons are available for Excel to use data spreadsheets.
- *Tableau:* This is good for anyone who wants to generate data stories on the web. Sharing the data is made through web and mobile. It extends support to all browsers across platforms.

- *Heap:* San Francisco-based Heap is a business intelligence provider. The BI tool Heap has prompted a host of intriguing business decisions and strategies.

The list is not exhaustive. For multidimensional analysis involving data warehouse, the most common tools include Microsoft SQL Server Analysis Services, SAP Business Objects, IBM Cognos, WebFOCUS, and Oracle Hyperion. There are many ways of outsourcing services in BI systems—reporting services, OLAP service, and data mining and visualization of data service [9].

## 14.6 Applications

Business intelligence has been widely adopted for over 30 years or so in the business world. Interest in implementation of business intelligence (BI) technologies in different sectors has been increasing from year to year. A typical deployment of a BI application is over the web (web search engines such as Google and Yahoo). BI data is typically stored in data warehouse. Its applications include the activities of decision support systems, reporting, on-line analytical processing (OLAP), statistical analysis, forecasting, and data mining. Although BI is designed for the business environment, it can be applied in various areas such as finance, retail, insurance, healthcare, government, universities, and security.

1. *Enterprise:* BI solutions are an indispensable component in the enterprise. BI may be regarded as a collection of decision support technologies for the enterprise aimed at enabling professionals make better and faster decisions. Companies or organizations that use BI are often large and multinational. BI is useful to startups, data-centric companies, and companies looking to grow rapidly. The business-driven approach to BI implementation is often motivated by competition and the need for gaining competitive advantage over business rivals. Companies can use BI to make a company's raw data usable. They apply BI to accelerate and improve decision-making, reduce costs, and identify new opportunities. BI blends and visualizes proprietary enterprise data to discover trends and patterns that can help businesses make better, data-driven, day-to-day decisions. Organizations that have implemented BI include Lenovo (the world's largest PC vendor), McDonald, MoneySuperMarket (MSM) in the UK, Coca-Cola Bottling Company, Chipotle, Des Moines Public Schools, Lowe's, Starbucks, and Amazon.
2. *Healthcare*: Today, healthcare industry is under pressure to improve operational efficiency, patient care, and economic sustainability. Healthcare practitioners are facing the demand for information to comply with legal and customer requirements. BI is considered a possible solution to this challenge. Hospitals have their data spread out and residing in several databases in the form of clinical data, administrative data, and external data [10]. BI system is developed to support decision-making domains such as hospital management, clinical decision support, and clinical research. Forward-thinking healthcare companies realize that

data and business intelligence play a crucial role in precise decision-making that will improve patient care. Healthcare is changing very fast and so is the industry's need for business intelligence. One may regard BI as a new methodology to maximize the benefits for healthcare organizations. Some applications of business intelligence in healthcare are shown in Fig. 14.4.

3. *Restaurant Chains*: Because restaurant chains such as McDonald, Wendy, and Hardee are operations-driven, they are heavy users of BI. They use BI to make timely business decisions and for tactical matters, such as what new menus to add, which dishes to remove, and which stores to close due to underperformance. In 2015, Uber launched its food-delivery app UberEATS. Since then, Uber has constantly collected business intelligence on restaurants. In 2016, Uber's engineers built Restaurant Manager, a scalable, user-friendly analytics tool that gave their partner restaurants access to real-time information [11].
4. *Mobile Devices*: BI no longer stays in one place like the office; it goes with us wherever we go. If your employees are on-the-move, support for mobile devices

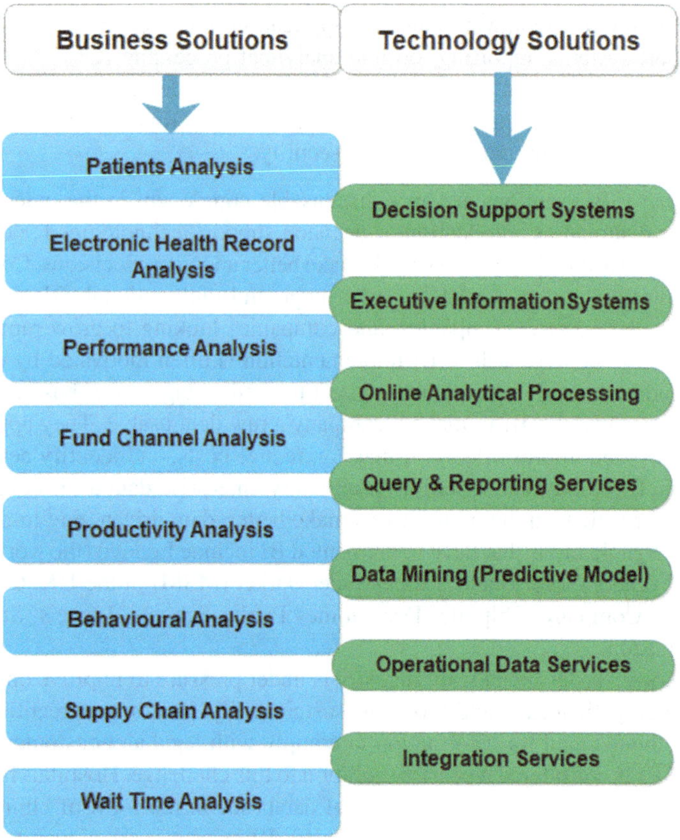


**Fig. 14.4** BI applications in healthcare
\

is critical. Developing BI application for the new generation of mobile devices (such as smart phones and tablets) is one of the major challenges facing practitioners [12]. The security of mobile BI systems is a major problem.

5. *Manufacturing:* A massive amount of data is managed in the manufacturing industry, with increased use of robotics and artificial intelligence. This makes the industry to be suited for utilizing business intelligence.
6. *Education:* Higher education plays a vital role in the socioeconomic development of any nation. Developing nations face some challenges on higher education programs. They experience problems such as financial insecurity, poor managerial practices, and system inefficiency. A business intelligence-based model can be used to integrate data from different sources to monitor the status of higher education as well as predict its future trends [13].
7. *Library:* Libraries are generally undergoing major changes due to information and technological revolution. Data plays a major role in helping academic library innovate their services and procedures accordingly. Business intelligence offers powerful analytical tools, such as visualization and data-mining tools, to transform data into information for decision-making and strategic planning for the library [14].
8. *Smart Grid*: The energy industry constantly seeks new ways to achieve higher energy efficiency in order to meet the ever-increasing demands of consumers. Achieving optimum energy efficiency is ensured by adopting of smart grid technologies. BI in smart grids maximizes the "smartness" of the grid. The use of BI systems in the energy industry has a potential to bring new value to business models [15].

Other areas of applications of business intelligence include technology, transportation, food and beverages, social platforms, e-commerce, retail sector, tourism industry, autonomous cars, crime prevention and solution, and emotional business intelligence.

## 14.7 Benefits

Business success largely depends on knowing the who, what, where, when, why, and how of the market and business intelligence can keep your organization in the know. The goal of business intelligence is to support better business decision-making. BI has brought us new possibilities on data analysis, data collection process, decision support systems as well as in financial, strategy, marketing, and production areas. It can eliminate a lot of the guesswork within an enterprise.

Business intelligence facilitates data-driven decision-making in an organization. BI tools are commonly used to make informed business decisions. They help on data visualization that enhances the data quality and decision-making. They provide full transparency, analysis, and delivery of financial and operational data. They enable faster, more accurate reporting and analysis, better employee satisfaction, reduced costs, and increased revenues. BI systems help businesses to identify market trends and spot business problems to be addressed. BI adoption helps organizations to manage the untapped potential of their massive amounts of data. Companies adopt BI for many reasons such as supporting functions as diverse as hiring, compliance, production, and marketing.

Other benefits of using business intelligence systems include [1, 16, 17]:

- *Decision-Making:* BI accelerates and improves healthcare decision-making. The faster data gathering and meaningful analytical report production helps in decision support. They also provide high-quality data for decision-making. BI systems also allow decision-makers to use and alter data and models in real time to support decision-making.
- *Protection of Data:* Since there would be a single point access of data; access is only provided to those with appropriate access authorization.
- *Increased Profits and Reduced Costs*: The need to reduce costs across all areas of operations is becoming necessary. BI allows organization to reduce waste and minimize the costs.
- *Prediction:* BI provides tools necessary to make accurate predictions. Making such prediction may require time, relevant data, sufficient information, and experience.
- *Boost Productivity:* BI tools save lots of time and resources, thereby allowing employees to be more productive.
- *Fix Accountability:* BI system assigns accountability in the organization as there must be someone who should own accountability and ownership for the organization's performance against its set goals.
- *It Streamlines Business Processes:* BI takes out all complexity associated with business processes. It also automates analytics by offering predictive analysis, computer modeling, benchmarking, and other methodologies.
- *It Allows for Easy Analytics:* BI software has democratized its usage by allowing nontechnical users to collect and process data quickly.
- *Increased ROI:* BI helps identify resources needed to reach goals and helps in the discovery of new revenue streams.

## 14.8 Challenges

The term "business intelligence" may turn out to be a fad. Since BI solutions often cover the whole spectrum of cost and features, evaluating all options and selecting the right one can be challenging. Adopting business intelligence in an organization faces several challenges [18, 19].

- *Cost:* Although most companies can benefit from incorporating BI solutions, BI system may be costly for small- and medium-sized enterprises. BI was first established for rich organizations. It is yet not affordable for many small- and medium-sized companies.

- *Lack of Funds:* Adopting and shifting to such a complex technology requires a lot of initial capital. Such a complex system requires availability of equipment like server, fast computing machines, and expensive software. Some organizations are challenged by limited budgets and ineffective management.
- *Complexity of IT Equipment:* Technically, some business intelligence systems are complex to implement and maintain; they find it difficult to keep up fast changing technologies. It may also be difficult to segregate between relevant and irrelevant data.
- *Lack of Knowledge:* Advanced technology requires the availability of the knowledgeable personnel who can help others to learn and use the technology appropriately and effectively.
- *Data Standards:* Organizations need to have a standard according to which such complex data can be structured. For example, interoperability is needed to allow sharing electronic health records with physicians, pharmacists, and hospitals.
- *Time-Consuming Implementation:* It takes almost one and half year for data warehousing system to be completely implemented. It is a time-consuming process.
- *Ethics:* Today, business ethics is gaining a great deal of attention in modern societies. There is the broad question of whether or not the collection and use of business intelligence is an ethical practice. Management must understand its primary duty to keep the organization's trade secrets safe by establishing and enforcing an effective security program [20].

## 14.9 Future of Business Intelligence

To learn from the past and predict the future, many companies worldwide are adopting BI tools and practices in their international business activities. Without doubt, BI technology is the future of business. The following are some business intelligence and analytics trends that will dictate the future of BI [21].

- *Artificial Intelligence:* AI and machine learning now take on complex tasks done by human intelligence. This capability is being leveraged to come up with real-time data analysis and dashboard reporting. Business intelligence can be described as the automated process to collect raw data from heterogeneous sources and organize them in a manner that insights can be derived from the data to improve business processes. Merging BI and AI will allow organizations to become more powerful, by selling services to each company and consumer.
- *Cloud Analytics:* Cloud computing offers new solutions for BI. BI applications will be soon offered in the cloud, and more businesses will be shifting to this technology. The reasons for adopting cloud computing technologies include improving agility and responsiveness, accelerating product development, and saving money. As enterprises move to the cloud, the BI system is automated using cloud-native BI applications that extract insights, make suggestions, and

create visual representations of the data. When choosing a cloud deployment, one should keep the issue of security in mind.
- *Collaborative BI:* BI software combined with collaboration tools, including social media, and other latest technologies enhance the working and sharing by teams for collaborative decision-making.
- *Embedded BI*: Embedded BI allows the integration of BI software or some of its features into another business application for enhancing and extending its reporting functionality.

The shifting trends will lead to next generation BI.

## 14.10 Conclusion

Business intelligence (BI) is an umbrella term for the technology that collects, stores, and analyzes the data produced by a company's activities. It helps derive meaningful insights from raw data. Since BI plays a key role in achieving and maintaining competitive advantage, it has attracted much attention from researchers and practitioners. The main objective of business intelligence is to enable business managers to have easy assess to data, be able to conduct analysis, allow them to convert data into useful knowledge, and then make faster, better decisions. An effective use of BI will result in time savings, cost rationalization, and management efficiency.

Although BI holds great promise, it faces technical and cultural challenges. The gap between deployed BI and BI users indicates that the full potential of BI is not fully tapped by companies and organizations. It has been reported that 70% of BI applications fail due to technology, cultural, and infrastructure issues [22].

Some academic institutions have started to offer courses on the concepts, techniques, and tools to provide effective business intelligence. More information on business intelligence can be found in the books in [2, 23–37] and the following related journals:

- *Business Intelligence Journal*
- *International Journal of Business Intelligence Research*
- *Journal of Decision Systems*

## References

1. J. Gubalova, The use of business intelligence tools for leadership and university administration, in *Proceedings of the 13th International Conference on Emerging eLearning Technologies and Applications (ICETA)* (2015), pp. 1–7
2. C. Vercellis, *Business Intelligence: Data Mining and Optimization for Decision Making* (Wiley, 2009)
3. M.N.O. Sadiku, M. Tembely, S.M. Musa, Toward better understanding of business intelligence. J. Scient. Eng. Res. **3**(5), 89–91 (2016)

4. Nikhilesh, Importance of business intelligence in healthcares (2012), https://www.helicaltech.com/blogs/importance-of-business-intelligence-in-healthcare/
5. C.M. Olszak, K. Batko, The use of business intelligence systems in healthcare organizations in Poland, in *Proceedings of the Federated Conference on Computer Science and Information Systems* (2012), pp. 969–976
6. R. Mulcahy, Business intelligence definition and solutions, http://aspnet.cob.ohio.edu/matta/MIS2020/HotTopics/3.0%20Business%20Intelligence.pdf
7. I. Hertz, Healthcare and BI: How can analytics improve patient care? (2018). https://technologyadvice.com/blog/healthcare/healthcare-bi/
8. M.S. Gounder, V.V. Iyer, A.A. Mazyad, A survey on business intelligence tools for university dashboard development, in *Proceedings of 3rd International Conference on Big Data and Smart City* (2016)
9. K. Rostek, Dedicated business intelligence system for SMEs consortium. Afr. J. Bus. Manag. **7**(13), 999–1014 (2013)
10. T. Mettler, V. Vimarlund, Understanding business intelligence in the context of healthcare. Health Informatics J. **15**(3), 254–264 (2009)
11. M. Rice, 15 Business intelligence examples and applications (2019), https://builtin.com/big-data/business-intelligence-examples-applications
12. P. O'Donnell, S. Sipsma, C. Watt, The critical issues facing business intelligence practitioners. J. Decis. Syst. **21**(3), 203–206 (2012)
13. V. Khatibi, A. Keramati, F. Shirazi, Deployment of a business intelligence model to evaluate Iranian national higher education. Soc. Sci. Human. Open **2**(1) (2020)
14. F. Hamad et al., *Business Intelligence in Academic Libraries in Jordan: Opportunities and Challenges* (Int. Feder. Lib. Assoc. Instit., 2020), pp. 1–14
15. M. Radenkovi et al., Harnessing business intelligence in smart grids: A case of the electricity market. Comput. Ind. **96**, 40–53 (2018)
16. The relevance of BI in healthcare industry, https://www.appsvolt.com/the-relevance-of-business-intelligence-in-healthcare-industry/
17. What is business intelligence? Definition and example, https://www.guru99.com/business-intelligence-definition-example.html
18. A. Khanna, D. Bhasin, Business intelligence in healthcare industry. Int. J. Sci. Res. **4**(12), 2136–2139 (2015)
19. N.O. Schultz et al., The ethics of business intelligence. J. Bus. Ethics **13**(4), 305–314 (1994)
20. D. de Carvalho et al., Business intelligence: future perspectives (2016), https://www.researchgate.net/publication/305649217_Business_Intelligence_Future_Perspectives_April_2016
21. C.M. Olszak, Toward better understanding and use of business intelligence in organizations. Inf. Syst. Manag. **33**(2), 105–123 (2016)
22. W. Grossmann, S. Rinderle-Ma, *Fundamentals of Business Intelligence* (Springer, 2015)
23. J. Khuntia, X. Ning, M. Tanniru, *Theory and Practice of Business Intelligence in Healthcare* (IGI Global, 2019)
24. B. Liautaud, M. Hammond, *e-Business Intelligence: Turning Information into Knowledge into Profit* (McGraw-Hill, 2000)
25. L.T. Moss, S. Atre, *Business Intelligence Roadmap: The Complete Project Lifecycle for Decision-Support Applications* (Addison-Wesley Professional, 2003)
26. S. Williams, N. Williams, *The Profit Impact of Business Intelligence* (Morgan Kaufmann, 2006)
27. D. Loshin, *Business Intelligence: The Savvy Manager's Guide*, 2nd edn. (Morgan Kaufmann, 2012)
28. B. Larson, *Delivering Business Intelligence with Microsoft SQL Server 2005* (McGraw-Hill Osborne Media, 2006)
29. V.L. Sauter, *Decision Support Systems for Business Intelligence*, 2nd edn. (Wiley, 2011)
30. E.R. Turban et al., *Business Intelligence*, 2nd edn. (PrenticeHall/Pearson, Englewood Cliffs, NJ, 2008)

31. L.B. Madsen, *Healthcare Business Intelligence: A Guide to Empowering Successful Data Reporting and Analytics* (Wiley, 2012)
32. S. Scheps, *Business Intelligence for Dummies* (For Dummies, 2011)
33. J. Machado, A. Abelha (eds.), *Applying Business Intelligence to Clinical and Healthcare Organizations (Advances in Bioinformatics and Biomedical Engineering)* (IGI Global, 2016)
34. C. Howson, *Successful Business Intelligence* (Tata McGraw-Hill Education, 2007)
35. Z. Michalewicz et al., *Adaptive Business Intelligence* (Springer, 2006)
36. R. Sharda, D. Delen, E. Turban, *Business Intelligence and Analytics: Systems for Decision Support*, 10th edn. (Pearson, 2014)
37. C. Howson, *Successful Business Intelligence*, 2nd edn. (McGraw-Hill Education, 2014)

# Chapter 15
# Augmented Intelligence

## Contents



> Some people call this artificial intelligence, but the reality is this technology will enhance us. So instead of artificial intelligence, I think we'll augment our intelligence.
>
> —Ginni Rometty

## 15.1 Introduction

Machines are performing better and better at analyzing data. Due to recent breakthroughs in technologies, organizations and businesses have started to experiment with the different ways that intelligent machines can assist humans. Leaders of organizations are increasingly under pressure to redesign their workflow processes and decide which tasks to *automate,* which tasks to *augment,* and which tasks to leave to humans. Today, artificial intelligence (AI) is shorthand for any task a machine can perform just as well as, if not better than, humans. AI is complementary to human intelligence and will not to take over human function [1].

There are many jobs that machines cannot do. As machines take over highly repeatable work, people will migrate to roles that call for critical thinking, creativity, judgment, and common sense. As shown in Fig. 15.1, augmented intelligence unites the strengths of humans and machines. One can augment human instinct with smart algorithms that provide fast, data-driven predictive insights. Augmented intelligence best combines human and artificial intelligence to change human behavior.


M. N. O. Sadiku, S. M. Musa, *A Primer on Multiple Intelligences*,
https://doi.org/10.1007/978-3-030-77584-1_15

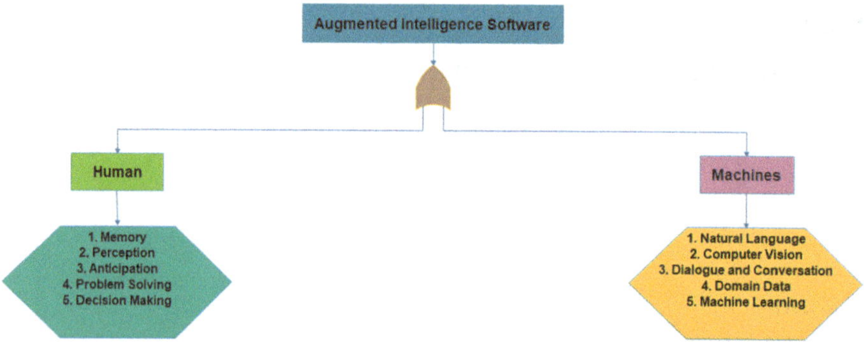


**Fig. 15.1** Augmented intelligence combines human and machine intelligences

It is one of the most important emerging technologies. It covers the subfields of AI, machine learning, and neural networks and it is going to disrupt the way in which companies and people work.

This chapter introduces the reader to the concept of augmented intelligence. It begins by describing what augmented intelligence is all about. It compares augmented intelligence with artificial intelligence. It provides some applications of augmented intelligence. It highlights the benefits and challenges of augmented intelligence. The last section concludes with comments.

## 15.2 What Is Augmented Intelligence?

Augmented intelligence (AI), also known as intelligence augmentation (IA) or cognitive augmentation, is next level in artificial intelligence. The word "augmented" means "to improve." Thus, the ultimate objective of augmented intelligence is using machines to enhance the work, expertise, and experience of a human user or worker. Augmented intelligence is also known as intelligence amplification (IA), cognitive augmentation, intelligence augmentation, machine augmented intelligence, or enhanced intelligence. It is essentially artificial intelligence (AI) with a twist [2]. AI software will simply improve products and services, not replace the humans that use them. Augmented intelligence is also AI. To avoid the confusion of using the same abbreviation for two meanings, let us call it IA: intelligence augmentation. IA refers the creation of a close-to-human autonomous intelligence using modern technology. It describes how normal human intelligence is supplemented through the use of technology. One may say that augmented intelligence results when man meets machine. In other words:

$$\text{Augmented Intelligence} = \text{Human} + \text{Computer}$$

One may think of IA as augmented reality, a technology which combines real-world environments with computer-generated information such as images, text, videos, animations, and sound. It has the ability to record and analyze the environment in real time. It is becoming more attractive as a mainstream technology mainly due to the proliferation of modern mobile computing devices like smartphones and tablet computers with location-based services. Like augmented reality, augmented intelligence adds layers of information on top of human intelligence, helping humans to be at their best.

The pioneers of augmentation are the sectors that generate a lot of data, such as the law, healthcare, and agriculture. The main objective of augmented intelligence is to create an entirely new process automated and designed for 20% manual exceptions.

## 15.3 How Augmented Intelligence Works

This is how augmented intelligence works. Augmented intelligence follows a five-function cadence that allows it to learn with human influence. As shown in Fig. 15.2, augmented intelligence repeats a cycle of understanding, interpretation, reasoning, learning, and assurance:

- *Understanding:* Systems are fed data, which they break down and derive meaning from.
- *Interpretation:* New data is inputted, the system then reflects on old data to interpret new data sets.
- *Reasoning:* The system creates "output" or "results" for new data set.
- *Learn:* Humans give feedback on output and the system adjusts accordingly. IA is an intelligence tool with human feedback loop.
- *Assure:* Security and compliance is ensured using blockchain or AI technology.

Having humans and machines work hand in hand is a win-win situation.

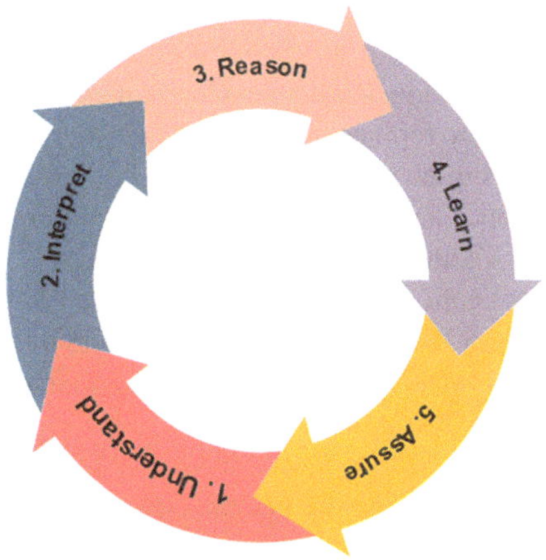


**Fig. 15.2** How augmented intelligence works

To effectively attain augmented intelligence, consider this checklist of five key areas [3]:

- Cultivate and look for the right talent, i.e., bilinguals with domain and technology expertise.
- Have proper governance over automation and people.
- Think through change management, specifically ensuring a smooth adoption in the new ways of work and upskilling talent.
- Make sure your technology and processes have continuous human input.
- Develop white-box algorithms that can explain why things are not working as they should.

Some companies are already focusing on developing smart data analytics solutions to obtain valuable insights from big data. Augmented analytics can be used to extract insights from big data. It automates data insights and provides clearer information, which is not possible with traditional analysis tools [4].

## 15.4 Comparing Artificial and Augmented Intelligence

It is important that we distinguish between AI and IA. In order to do that, it is expedient to understand what AI is. Artificial Intelligence (AI), as its name implies, is a different form of intelligence to our own. It is human-like intelligence that works in a similar way to our brains. Artificial intelligence (AI) is the cognitive science that deals with intelligent machines which are able to perform tasks heretofore only performed by human beings. It is essentially the simulation of human intelligence by machines. Artificial intelligence is everywhere: from wearable products to driverless cars. Even the system that prevents a car from starting because a door is open is artificial intelligence in action [5].

AI and IA are some of the disruptive technologies that are affecting organizations, economies, and our societies. Although augmented intelligence (IA) and artificial intelligence (AI) are closely related, they are different. The main difference between them is one of autonomy. AI operates without human interaction. IA uses AI techniques but keeps the human in the loop. The goal of augmented intelligence is not to replace human activities, but instead to elevate existing human capabilities. AI is often designed to mimic human intelligence, while IA enhances human intelligence and makes it work faster and smarter.

Another key difference between AI and IA lies in the cognitive learning and application of any machine. Unlike AI, which tries to replicate human intelligence, IA works with and amplifies human intelligence.

Augmented intelligence is a subfield of AI developed to enhance human intelligence. It uses some AI tools such as machine learning and deep learning to provide humans with actionable data. Augmented intelligence tools are created to help rather than replace humans. Augmented intelligence focuses on software that takes over small repetitive tasks. It combines human and machine intelligence. Focusing just on AI can expose companies to unforeseeable challenges in the future. Instead of fearing artificial intelligence as replacement technology, we should embrace augmented intelligence and find ways we can use it to alleviate the fears.

While AI is meant to help us humans with things like computation, memory, perseverance, precision, recognition, and speed, augmented intelligence allows us to see how it can help augment more human things like abstraction, breaking rules, judgment, listening, and storytelling [6]. Augmented intelligence, not artificial intelligence, is the future. Therefore, businesses should be focusing on augmented intelligence instead of AI.

## 15.5 Applications

Augmented intelligence is a tool for finding hidden meaning within data by uncovering patterns and correlations at a massive scale. Interest in augmented intelligence (AI) is growing rapidly among Congress and federal agencies. Augmented intelligence can be applied in business, decision-making, healthcare, journalism, insurance underwriting, ID verification, logistics planning, remote assistance, manufacturing, finance, real estate, military, and legal profession.

- *Business Sales*: The merger of humans and machines is critical to advancing modern business enterprise. Departmental heads, team leads, and vice presidents often identify business opportunities to apply augmented intelligence. Although augmented intelligence systems can be placed anywhere in a business organization, it is mostly used in the business sales. Integrating intelligent capabilities into your sales department is a smart investment in your salespeople. Augmented intelligence takes the complex sales tasks off the salespersons and enables them to focus on meeting the needs of customers. AI is capable of analyzing sales data and translate the data into action [7]. The range of business problems to which IA applies continues to expand at a rapid pace.
- *Healthcare:* Patients and healthcare practitioners face enormous challenges: rapidly aging population, shortage of physicians, high costs of care, etc. This is partly illustrated by the quadruple aim of healthcare shown in Fig. 15.3. Augmented intelligence has the potential to transform our experience of health, healthcare, and wellness. It can offer a transformative set of tools to help patients, physicians, and the nation face these looming challenges [8]. Augmented intelligence is where clinical knowledge and medical data converge on a single platform. The primary role for AI is augmentation of the intelligence and skills of caregivers. IA is not designed to replace healthcare practitioners but to enhance human intelligence and the physician/patient relationship. The healthcare industry is facing major disruption as market and regulatory forces drive an even greater need for operational efficiency while delivering better service.
- *Decision-making*: If properly applied, augmented intelligence can provide feedback and insights that enhance decision-making. Augmented intelligence

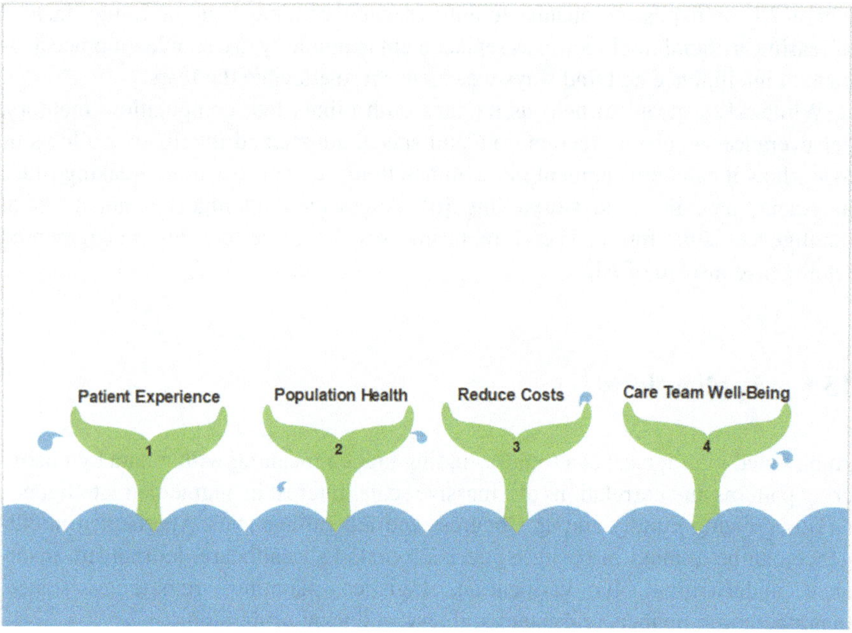


**Fig. 15.3** The quadruple aim of healthcare

involves a creative mix of data, analytics, artificial intelligence (AI), and human judgment. It is the ability of a manager to leverage artificial intelligence and collective intelligence for every decision. It guides human employees to make smarter decisions and find new discoveries. The future of decision-making requires a transition into a whole new world and way of thinking [1].

- *Logistics Planning:* A logistics planner requires a lot of knowledge and experience about what works and what does not work in the industry. Augmented intelligence comes in when artificial intelligence is used to deal with high-skilled logistics planners. To improve logistics planning, companies should use augmented intelligence, which combines inputs from human planners with artificial intelligence technology [9].
- *Remote Assistance:* Current remote systems rely on human intervention. Augmented intelligence and machine learning offer the potential to support remote assistance. This may be a situation where a field operator requires the technical support of an expert. Augmented intelligence is employed to assist human memory in the recall, recognition, and search of information in real time [10].
- *Military*: The military seeks to develop augmented intelligence that combines the strengths of humans and machines to create a human edge in the information age. Combinations of humans and machines are necessary to cope with the complexity and maximize the military ability to create, exploit, and adapt. Augmented

intelligence will harness the best of our soldiers and technology to meet future challenges [11].

- *Education*: As we move from the information age into the creative age, education at all levels will need to change and adapt accordingly. A key challenge for educators is understanding the critical importance of human creativity in the development of artificial intelligence and intelligence augmentation. The biggest challenge is to have efficient access to knowledge. The knowledge gap is real and it cannot be fought with the tools of the past.

Other applications of augmented intelligence include financial services, retail, travel, manufacturing, and oil and gas.

## 15.6 Benefits

Augmented intelligence, the synergy between humans and artificial intelligence, exploits the positive aspects of human and AI-generated reasoning. It is designed to play an assistive role to enhance human intelligence and to improve the quality and efficiency of the workers. It is becoming an effective weapon to address the current blemishes of medicine which include poor predictive power, therapeutic errors, and inefficient hospital workflows [12]. IA tools are used in a number of fields to help drive productivity, improve efficiency, make smarter decisions, waste less time on repetitive tasks, and save organization time and money.

IA is better for organizations than AI as it has a quicker ROI and better utilizes the assets already in place. There are opportunities that will arise from "augmented" workplaces and life in general. IA transforms social services organizations from case management to care management. Any field that requires both humans and technology will definitely be changed by augmented intelligence.

Augmented intelligence offers data scientists a tool to sift through big data. Without augmented intelligence, processing a vast amount of data would take a long time and data scientists would be faced with a huge problem. Thus, augmented intelligence has applications in any industry that mines big data for analytics and predictive indicators. It improves human decision-making by handling big data that would overwhelm a human.

## 15.7 Challenges

While there are benefits to using AI, there are several legal and ethical issues. There are serious issues concerning customer privacy and security. The idea that intelligence can be automated, replacing millions of humans in highly repeatable tasks, has received a great attention. For the doomsayers, AI-augmented workplace implies an impending mass unemployment. They fear that many white-collar professionals risk being automated, from taxi drivers to telemarketers and receptionists. The fears are essentially groundless. It appears that jobs will disappear rather than being automated. If machines can replace humans, we should free to do what humans do best: create.

## 15.8 Conclusion

Intelligent machines (or computers) are efficient in numerical computation, information retrieval, statistical reasoning, and have almost unlimited storage. They can see, hear, and speak multiple languages. They have become the intimate companions of humans; they are profoundly changing our lives and shaping the future. We are in an era where machines and humans working together will be pervasive and efficient.

The human brain is an extremely complex advanced processing unit that can solve problems that computers cannot solve. No matter how intelligent machines may be, they are cannot completely replace humans. So there is nothing to fear. We only need to prepare. In the foreseeable future, AI will be augmenting our capabilities, allowing us to do more in less amount of time. Rather than focusing on AI, we should consider the possibilities of how businesses can take advantage of augmented intelligence. For more information on augmented intelligence, one should consult the books in [13–17].

## References

1. M.N.O. Sadiku, S.M. Musa, A. Ajayi-Majebi, Ambient intelligence: a primer. Int. J. Trend Res. Develop. **7**(5), 71–74 (2020)
2. What is augmented intelligence and why should you know about it? Unknown Source
3. The future of decision-making: augmented intelligence, https://www.genpact.com/insight/point-of-view/the-future-of-decision-making-augmented-intelligence
4. Augmented analytics plays a important role in transform big data to smart data (2019), https://timestech.in/augmented-analytics-plays-a-important-role-in-transform-big-data-to-smart-data/
5. M.N.O. Sadiku, Artificial intelligence, in *IEEE Potentials* (1989), pp. 35–39
6. What is augmented intelligence? (2018), https://www.mediaupdate.co.za/media/143606/what-is-augmented-intelligence
7. S. Akhrif, Augmented intelligence and the future of selling, https://uib.ai/augmented-intelligence-future-selling/
8. Health imaging: augmented intelligence the next frontier, https://www.healthcareitnews.com/news/augmented-intelligence-next-frontier-health-imaging
9. Augmented intelligence for logistics planning, https://transmetrics.eu/blog/augmented-intelligence-for-logistics-planning/
10. W.L. Koh, et al, Preliminary investigation of augmented intelligence for remote assistance using a wearable display, in *Proceedings of the 2017 IEEE Region 10 Conference*, Malaysia (2017), pp. 2093–2098

11. Augmented intelligence, https://www.airforce.gov.au/our-mission/plan-jericho/augmented-intelligence
12. M. Bhandari, M. Reddiboina, Augmented intelligence: a synergy between man and the machine. Indian J. Urol. **35**(2), 89–91 (2019)
13. A. Bates, *Augmented Mind: AI, Humans and the Superhuman Revolution* (Neocortex Press, 2019)
14. D. Araya (ed.), *Augmented Intelligence: Smart Systems and the Future of Work and Learning* (Peter Lang Inc., International Academic Publishers, 2018)
15. M. Pasquinelli (ed.), *Alleys of Your Mind: Augmented Intelligence and Its Traumas* (Meson Press, 2015)
16. J. Hurwitz et al., *Augmented Intelligence: The Business Power of Human–Machine Collaboration* (Auerbach Publications, 2019)
17. B. Larson, *Delivering Business Intelligence with Microsoft SQL Server*, 4th edn. (McGraw-Hill Education, 2016)

# Chapter 16
# Abstract Intelligence

**Contents**



> Our vanity, our passions, our spirit of imitation, our abstract intelligence, our habits have long been at work, and it is the task of art to undo this work of theirs, making us travel back in the direction from which we have come to the depths where what has really existed lies unknown within us.
>
> —Marcel Proust

## 16.1 Introduction

Intelligence has been a key component of modern psychology, the science of human behavior. It is an ability to acquire and use knowledge. It is regarded as a complex of mutually independent skills. It is a form of driving force that transfers information into actions in a human or a system. It plays a crucial role in cognitive informatics, computing, software science, brain science, and knowledge science [1]. Intelligence may also be regarded as the ability to learn from past experience and adapt to suit the environment. It is the combination of abilities required for survival and advancement in a particular culture or environment. Some characteristics of intelligence are shown in Fig. 16.1.

There is no one overall conception of intelligence. Howard Gardener, an American developmental psychologist, described nine types of intelligence in 1983 [2]:


M. N. O. Sadiku, S. M. Musa, *A Primer on Multiple Intelligences*,
https://doi.org/10.1007/978-3-030-77584-1_16

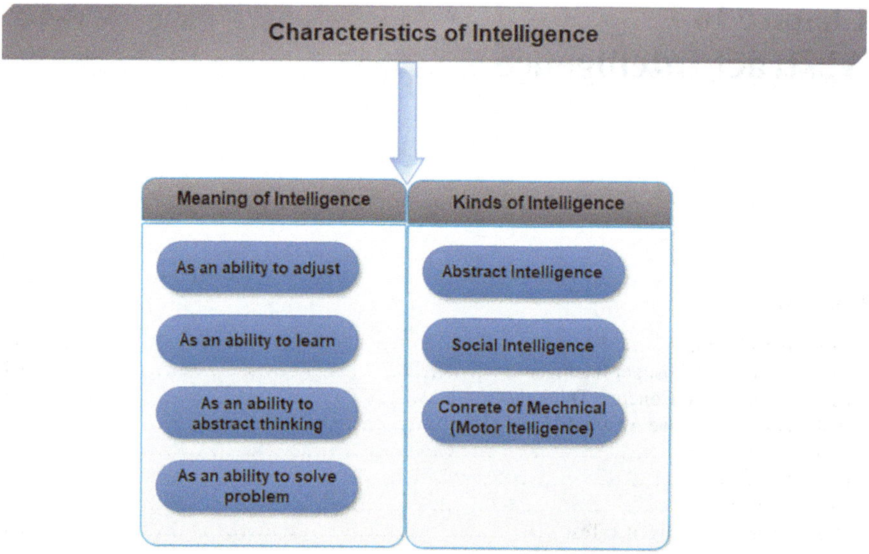


**Fig. 16.1** Characteristics of intelligence

- Naturalist (nature smart)
- Musical (sound smart)
- Logical-mathematical (number/reasoning smart)
- Existential (life smart)
- Interpersonal (people smart)
- Bodily-kinesthetic (body smart)
- Linguistic (word smart)
- Intrapersonal (self smart)
- Spatial (picture smart)

These are illustrated in Fig. 16.2.

As shown in Fig. 16.3, an American psychologist Edward Lee Thorndike also divided intelligence into three types:

1. Social intelligence or ability to understand and deal with persons
2. Concrete intelligence or ability to understand and deal with things, as in skilled trades and scientific appliances
3. Abstract intelligence or ability to understand and deal with verbal and mathematical symbols

Other types of intelligence not in the above lists include military intelligence, emotional intelligence, spiritual intelligence, swarm intelligence, machine intelligence, artificial intelligence, computational intelligence, digital intelligence, and augmented intelligence. Some forms of intelligence can be measured by standardized tests. It has been suggested that various paradigms of intelligence, such as natural, artificial, and computational intelligence, can be combined at the logical and

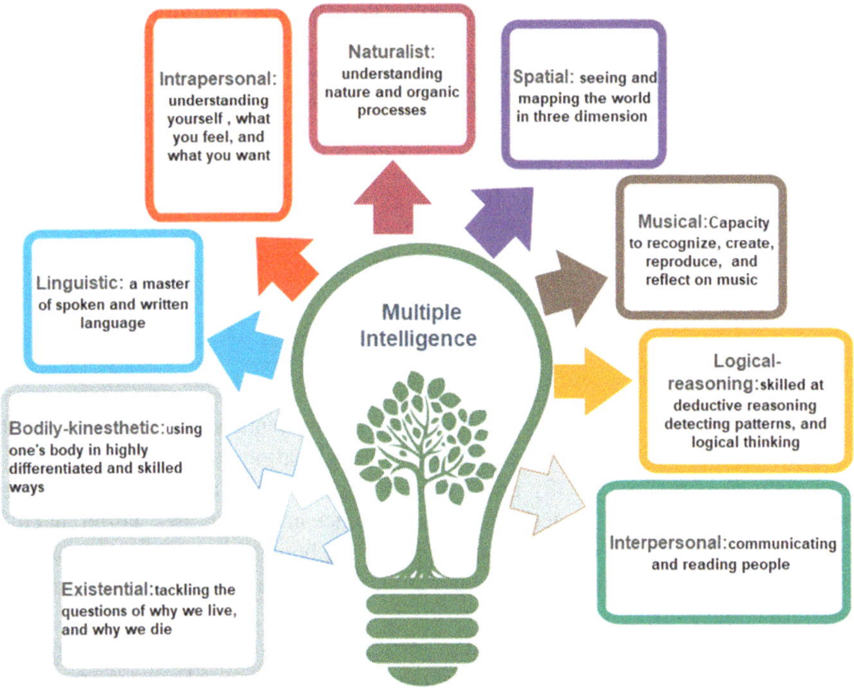


**Fig. 16.2** Nine types of intelligence

**Fig. 16.3** Three types of intelligence

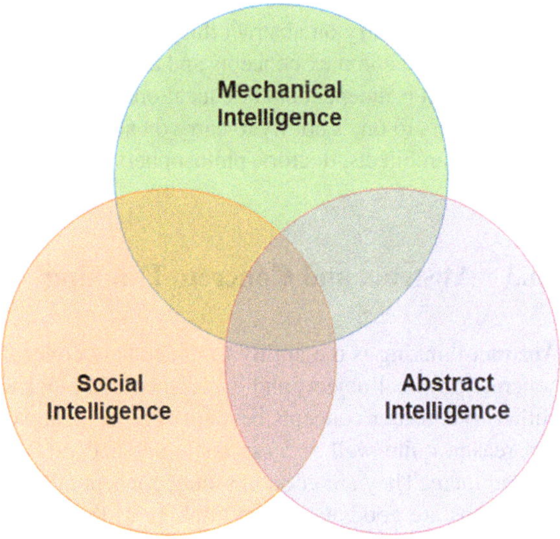

functional levels to form abstract intelligence, which is the ability to engage in abstract thinking [3].

This chapter introduces the reader to abstract thinking and abstract intelligence. It begins by describing what abstract intelligence is all about. It compares abstract and concrete thinking. It provides how one can improve abstract intelligence. It covers some applications of abstract intelligence. It highlights some benefits and challenges of abstract intelligence. It gives a simple test for measuring abstract intelligence. The last section concludes with comments.

## 16.2 Concept of Abstract Intelligence

Intelligence is being studied by psychologists, neuroscientists, and many others. It is affected by culture, environment, heredity, and education.

Intelligence cannot be fully understood outside its cultural context. There are substantial differences in conceptions of intelligence across cultures. Different cultures have different conceptions of intelligence and how it is manifested in day-to-day living. Some cultures emphasize the social aspect of intelligence than do Western cultures [4].

Intelligence may be regarded as the capacity for abstraction. Abstract intelligence has been introduced in order to explain the complex mechanisms of natural and artificial intelligence and as a human enquiry of cognitive informatics and intelligence science.

Abstract intelligence is a form of driving force that transfers information into behaviors or actions [5]. It is the ability to respond to words, numbers, letters, etc. It is the ability to carry on abstract thinking. It is a measure of one's ability to reason and understand complex concepts and assimilate new information beyond previous experience. It is independent of educational and cultural background. Abstract intelligence refers to one's ability to carry on abstract thinking. Good teachers, lawyers, engineers, architects, doctors, philosophers, etc. have this type of intelligence.

## 16.3 Abstract and Concrete Thinking

Abstract thinking is the ability to understand concepts that are real but not tied to concrete physical objects and experiences. In other words, abstract thinking is the ability to consider concepts beyond what we observe physically. Abstract thinkers can reason quite well and use analogies and metaphors to understand the world around them. They are good in seeing conceptually what is not there. For example, comedians are good at abstract thinking as they make jokes out of the unexpected connections. Math skills are often abstract as they depend on the ability to conceptualize things without laying hands-on physical objects. The study of science, language, history, social studies, philosophy, and politics all require the ability to think generally and express abstract ideas [5, 6].

The opposite of abstract thinking is concrete thinking, which is tied to objects and experiences that are observable. The terms "concrete" and "abstract" are used to suggest how practical or impractical an idea may sound. Abstract ideas are usually invisible, complex, and subjective, while concrete ideas are often visible and objective. Examples of abstract concepts include love, intelligence, bravery, happiness, loyalty, and convenience. Examples of concrete ideas include bank manager, college student, dog, computer, and pencil. A concrete thinker focuses on what is physically around him. Pure mathematics deals with abstract concepts, while applied mathematics deals with concrete concepts. For example, a concrete thinker may take everything too literally. The ability to think abstractly develops as we grow. Children begin to think like concrete thinkers. They observe and use their five senses and motor skills. As the child grow, their way of thinking changes and they gradually become abstract thinker at the last stage of their mental development [7].

Abstract reasoning questions can serve as a good measure of general intelligence. An abstract reasoning test is becoming popular as part of any job assessment or intelligence testing setup because they are independent of cultural background, education, and language. It measures one's nonverbal ability to reason logically and identify and interpret patterns. It is designed to assess the candidate's ability to work out new concepts and abstract ideas. It has been found to have a high correlation with general intelligence and the ability to reason logically [8]. We all need to use a blend of concrete and abstract thinking to function well daily.

Besides abstract and concrete thinking, there are other forms of thinking including linear thinking, nonlinear thinking, analytical thinking, creative thinking, critical thinking, divergent thinking, and convergent thinking. Abstract intelligence triggers the recent breakthroughs in cognitive systems such as cognitive computing and cognitive learning. Cognitive computing attempts to enable the computer to do what can only be understood and computed by human intelligence [9].

## 16.4 How to Improve Abstract Thinking

Even those who have mental impairments will have some abstract thinking. People are at different part of the spectrum. An average individual could learn how to think abstractly and grow. If someone is struggling with abstract thinking, some therapists can help them become better at abstract thinking. It is essentially to learn how to become a more abstract thinker if you want to improve your life.

If you want to improve your abstract thinking skills and abstract intelligence, try the following things [10]:

- *Improvise:* Consider taking a workshop that helps you to explore this open-ended form of performance play.
- *Solve puzzles:*3D, visual, and word puzzles will train you to think of alternatives

- *Build 3D models:* Research shows that people in science, technology, engineering, and mathematics professions utilize their abstract thinking abilities by working on different projects.
- *Play with figurative language:* The ability to write similes, metaphors, analogies, and even pieces of personification can stimulate abstract thinking.

Reading books aloud to children also contributes to the development of their abstract intelligence. Books expand the imagination of children and greatly help in the development of their abstract intelligence. As the children grow, so does their way of thinking.

## 16.5 Applications

Intelligence is regarded as a mental trait, innate capacity to solve the problem in the light of past experience and knowledge. It is an ability to carry on abstract thinking. Abstract intelligence is required in the most academic subjects in the school. It is demonstrated in career/profession, school, cognitive systems, and abstract language.

- *Profession*: Our profession is associated with intelligence. Although everyone possesses abstract intelligence to some degree, some people (such as engineers, mathematics, artists, designers, and architects) apply this intelligence on a regular basis. Comedians are experts in abstract thinking and in utilizing abstract intelligence. They observe the world around them and create jokes out of the unexpected connections. People with high abstract intelligence are often good at solving complex problems, taking intelligence tests, and coming up with novel ideas. Who do we considered to be more intelligent? An engineer designing an aircraft, a manager motivating his employees, a professor teaching a class on Chemistry, a violin player in a symphony, an author writing a book on religion?
- *School:* One of the most vital factors that affect schooling is abstract intelligence. Some of the tasks students perform in school are related to abstract intelligence. Math skills are often abstract and depend on being able to conceptualize numbers and operations. Science requires students to propose, test, and revise hypotheses and theories. The study of language, history, social studies, philosophy, and politics requires the ability to think generally and arrive at conclusions [11].
- *Cognitive Systems:* Basic studies in denotational mathematics and mathematical engineering have led to the theory of abstract intelligence. Abstract intelligence triggers the recent breakthroughs in cognitive systems such as cognitive computers, cognitive robots, cognitive neural networks, cognitive learning, and brain informatics. Figure 16.4 illustrates the theoretical framework of intelligence sciences.
- *Abstract language:* This includes indirect uses of language, such as metaphors and figures of speech. An abstract thinker would understand that the figure of speech means that people who have faults of their own should not criticize others. A person can improve performance with brain teasers, math problems, and

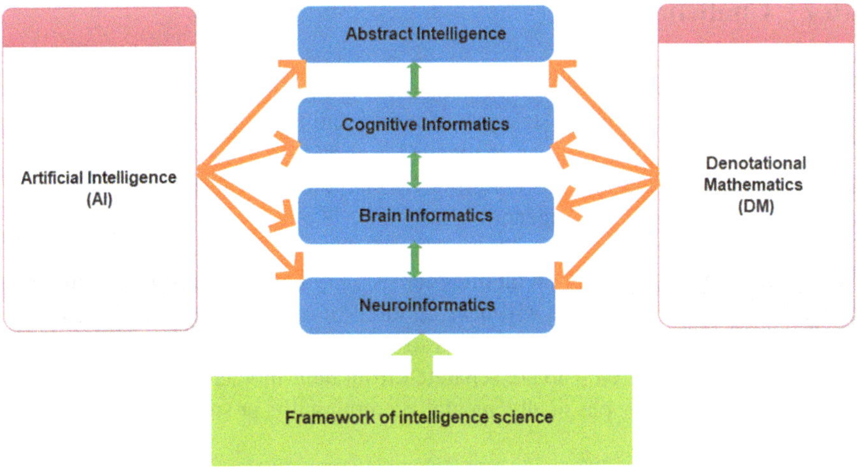


**Fig. 16.4** The framework of intelligence sciences

logic problems with no transfer to other domains of thinking. Therefore, attempts to facilitate increasingly abstract thinking should be made within all relevant academic areas (e.g., math, literature, science, social studies). If possible, similar language and analogies should be used (e.g., by parents and teachers) across areas so as not to overwhelm students with too much information [12].

Other applications of abstract intelligence include brain informatics and abstract artificial intelligence.

## 16.6 Benefits

It is crucial to have abstract thinking skills to navigate your way through life. Without abstract thinking, most of our inventions today would not exist. Abstract reasoning can benefit students in the classroom and in the real world alike. Some people such as mathematicians, artists, comedians, writers, and engineers utilize this abstract thinking a regular basis. Children can develop their abstract reasoning skills using blocks, tinker toys, geometric toys, or other items.

Abstract intelligence gave birth to cognitive systems such as cognitive computers, cognitive robots, cognitive neural networks, cognitive learning, cognitive informatics, cognitive robots, cognitive learning engines, cognitive engineering, and cognitive self-driving cars [11]. Abstract intelligence is generally found in "bookish" persons. They are critical thinkers, science-minded, and can reason very well. Abstract thinkers can do better on intelligence tests. They are the ones who produce and make lasting impact on the society. Some people may not understand abstract thinkers, but they can enjoy their contributions.

## 16.7 Challenges

There is no general definition of intelligence. Although people talk about intelligence, they find it difficult to state just what it is and how it is to be measured. The importance of IQ scores has been challenged over and over in the recent times. This has encouraged researchers to conceptualize alternative abilities such as social, abstract, and emotional intelligence.

Since heredity determines the limit of an individual's intelligence, change cannot be expected. Due to this, certain families consistently will have offspring with superior ability and other families will have children of inferior ability. Children from impoverished economic background fail to achieve success in elementary educational curricula. Children who are separated from their mothers soon after birth do not develop adequately physically, mentally, emotionally, or socially [13].

## 16.8 Abstract Intelligence Test

A number of tests have been developed to measure abstract intelligence. A simple test is presented in this section [14]. Circle the item in each series that does NOT belong with the other three, then in the space provided explain why the items does not belong. For example: in the series of cat, dog, mouse, and apple, apple does not belong because it is a fruit while the others are all animals.

NOTE: One point for correct answer and two points for the correct explanation. There is only one correct answer for each question. All other answers are wrong.

| | | | | |
|---|---|---|---|---|
| 1. | 2 | 8 | 7 | 4 |
| 2. | Baseball | Golf | Basketball | Football |
| 3. | Chan | Sawyer | Hardy | Holmes |
| 4. | President | Chief justice | Senator | Governor |
| 5. | 36 | 72 | 121 | 144 |
| 6. | Father | Son | Mother | Child |
| 7. | Indifference | Love | Hate | Concern |
| 8. | Slice | Hook | Curve | Top |
| 9. | Potato | Carrot | Squash | Onions |
| 10. | 9 | 7 | 4 | 8 |
| 11. | Harold | Sam | Remus | Tom |
| 12. | Whale | Man | Bass | Porpoise |
| 13. | A | H | C | L |
| 14. | Tomato | Pea | Corn | Lettuce |
| 16. | Peanut | Almond | Macadamia | Cashew |
| 16. | Good | Justice | Evil | Law |
| 17. | Axial | Radial | Distal | Proxial |

| | | | | |
|---|---|---|---|---|
| 18. | Limestone | Shale | Sandstone | Marble |
| 19. | White | Dwarf | Red sun | Moon |
| 20. | Isostatic | Fault | Complaint | Fissure |

| | |
|---|---|
| 55–60 very gifted | 25–28 average |
| 49–54 very superior | 19–24 low average |
| 42–48 superior x | 15–18 below average |
| 36–41 above average | 00–14 poor |
| 29–35 high average | |

Culture can influence a child's performance on testing so that the performance of Euro-American and African-American children differ.

## 16.9 Conclusion

The brain is a complex organ and no one has the same way of thinking. Intelligence is a global concept that involves a person's ability to act purposefully, think rationally, and deal effectively with the environment. It is the capacity to acquire and apply knowledge. More information on abstract intelligence can be found in the books in [14, 15].

## References


1. Y. Wang, On abstract intelligence: Toward a unifying theory of natural, artificial, machinable, and computational intelligence. Int. J. Soft. Sci. Comput. Intell. **1**(1), 1–17 (2009)
2. M. Vital, 9-types-of-intelligence – Infographic, https://blog.adioma.com/9-types-of-intelligence-infographic/
3. Y. Wang, On abstract intelligence and its denotational mathematics foundations, in *Proceedings of the* 2008 7th IEEE International Conference on Cognitive Informatics (2008)
4. R.J. Sternberg, Intelligence. Wires Cogn. Sci. **3**(5), 501–511 (2012)
5. R. Joy, Abstract thinking: what it is, why we need it, and when to rein it in (2019), https://www.healthline.com/health/abstract-thinking
6. M.N.O. Sadiku, O.D. Olaleye, S.M. Musa, Abstract intelligence: An introduction. Int. J. Trend Res. Develop. **7**(3), 21–23 (2020)
7. J. Thomas, The difference between concrete vs. abstract thinking (2020), https://www.betterhelp.com/advice/self-esteem/the-difference-between-concrete-vs-abstract-thinking/
8. Abstract reasoning test, https://www.123test.com/abstract-reasoning-test/
9. Y. Wang et al., Abstract intelligence: embodying and enabling cognitive systems by mathematical engineering. Int. J. Cogn. Inform. Nat. Intell. **11**(1), 1–22 (2017)
10. Abstract thinking: what it is, why we need it, and when to rein it in, https://www.healthline.com/health/abstract-thinking#takeaway

11. Tutorial: concrete vs. abstract thinking, http://www.projectlearnet.org/tutorials/concrete_vs_abstract_thinking.html#:~:text=Abstract%20thinkers%20are%20able%20to,attributes%20or%20share%20those%20relationships.&text=A%20concrete%20thinker%20can%20count,thinker%20can%20think%20about%20numbers
12. S. Gautam, Intelligence: definitions and characteristics, Term paper, Psychology, https://www.psychologydiscussion.net/term-paper/intelligence-term-paper/intelligence-definitions-and-characteristics-term-paper-psychology/13524
13. Abstract intelligence test, https://docs.google.com/viewer?a=v&pid=sites&srcid=c21pdGh2aWxsZS5rMTIubW8udXN8YXAtcHN5Y2hvbG9neXxneDo4MDE2NWYwMmRkODNlNzM
14. I. Yardley, *Abstract Intelligence: How to Survive, and, Excel, in Digital Reality* (CreateSpace Independent Publishing Platform, 2017)
15. Y. Wang, *Advances in Abstract Intelligence and Soft Computing* (IGI Global, 2012)

# Chapter 17
# Swarm Intelligence

## Contents



> The most amazing thing about social insect colonies is that there's no individual in charge. If you look at a single ant, you may have the impression that it is behaving, if not randomly, at least not in synchrony with the rest of the colony. You feel that it is doing its own things without paying too much attention to what the others are doing.
>
> —Eric Bonabeau

## 17.1 Introduction

Advancement of technology has led to problems that are complex and more challenging. Swarm intelligence techniques were mostly developed for solving optimization problems. The idea of swarm intelligence comes from swarming behaviors of groups of organisms. A swarm consists of a large number of simple entities that interact with each other and with the environment with no central control. The concept of swarm intelligence (SI) is an emerging domain which belongs to the field of artificial intelligence (AI), as many of its pursuits can be linked to machine learning.

Swarm intelligence is the emergent collective intelligence of groups of simple agents. It is the collective behavior of decentralized, self-organized systems, natural or artificial. It belongs to the emerging field of bio-inspired soft computing and optimization. It is inspired from the biological entities such as birds, fish, ants, wasps, termites, and bees. Bio-inspired computation is a field of study that is closely related to artificial intelligence [1]. Figure 17.1 shows bees' comb, a nature example of swarm intelligence [2].



M. N. O. Sadiku, S. M. Musa, *A Primer on Multiple Intelligences*,
https://doi.org/10.1007/978-3-030-77584-1_17

**Fig. 17.1** Bees' comb: Nature example of swarm intelligence [2]

Swarm intelligence is a complex discipline that seeks to understand and replicate both natural and artificial systems that are composed of several parts but effectively coordinate their activities through decentralized control. This type of intelligence is often displayed in nature through the habits of animals like bees, fish, birds, and ants. These animals can share information among themselves, giving them a great survival advantage. Thus swarm intelligence can be regarded as the collective behavior that emerges from social insects working together and following few rules. Swarm intelligence is the collective behavior emerging in systems with locally interacting entities. It enables individuals or entities to converge and create an organism that can do things that individuals cannot do by themselves.

This chapter provides an introduction to swarm intelligence. It begins by discussing what swarm intelligence is all about. It discusses SI system and techniques. It presents some applications of SI. It highlights the benefits and challenges of SI. The last section concludes with comments.

## 17.2 What Is Swarm Intelligence?

Mother Nature provides inspiration for naturalists, biologists, computer scientists, and other researchers in the most amazing ways. Swarm intelligence is a good example of this. Swarm intelligence (SI) is an emerging field of biologically inspired artificial intelligence technique based on the study of cooperation of simple individuals, such as ant colonies, bird flocking, animal herding, and bees, interacting locally with one another and with their environment. It is the collective intelligence of groups of simple autonomous agents in decentralized, self-organized environments. A swarm may be regarded as a large number of homogenous (or heterogeneous) individuals that have voluntarily converged for a common goal. Their inspiration usually comes from nature or biological systems.

Swarm intelligence was introduced by Gerardo Beni and Jing Wang in 1989 with their study of cellular robotic systems. The scientific discipline was born from biological insights about the incredible abilities of social insects to solve their everyday-life problems. Swarm intelligence is often regarded as a branch of evolutionary computation, which is a family of population-based optimization techniques which have a potential global search ability.

Recent studies have shown that individual social insects such as ants or bees follow simple rules and have no centralized control. The collective behavior of the swarm presents an intelligence that is beyond the intelligence of each individual member. In other words, a group of bees works better than one alone. In the bee colony, each bee has its own division of labor and unique work. A human swarm is a social network of individuals behaving for a time like-mindedly. The more sophisticated the brain, the more sophisticated the swarm behavior [3]. Some researchers even consider the collective, the swarm, as one single and unique individual.

Examples of swarm intelligence in natural systems include colonies of ants and termites, schools of fish, flocks of birds, herds of animals, rookeries of penguins, plagues of locusts, hawks hunting, and bacterial growth. The flocking and schooling are examples of highly coordinated group behaviors, where no leader is in charge and each individual bases its movement decisions just on locally available information. Human swarm is formed by a group of networked distributed human participants, thinking together as a real-time system and moderated by swarming algorithms. Such real-time systems enable the participants to behave as a unified collective intelligence that works as a single entity to make predictions [4]. Some human artifacts also fall into the domain of swarm intelligence, notably some multi-robot systems, and computer algorithms for tackling optimization problems.

The swarm principles are illustrated in Fig. 17.2 and explained as follows [5]:

*Principle #1: Awareness*. Each member must be aware of the surroundings and abilities.

*Principle #2: Autonomy.* Each member must operate as an autonomous master (not as a slave); this is essential to self-coordinate allocation of labor.

*Principle #3: Solidarity.* Cooperation is needed when a team achieves a task that could not be done by a single individual. Each member must cooperate in solidarity. When a task is completed, each member should autonomously look for a new task.

**Fig. 17.2** The five principles of swarm intelligence [5]

*Principle #4: Expandability.* The system must permit expansion where members are dynamically aggregated.

*Principle #5: Resiliency* The system must be self-healing; when members are removed, the remaining members should undertake the unfinished tasks.

## 17.3 SI Systems

Swarm intelligence is a biological phenomenon in which groups of organisms or participants amplify their combined brainpower by forming real-time systems. Swarming creates a unified system in which participants work together in real time. A swarm intelligence system typically consists of a population of simple agents which can interact or communicate with each other and with their environment and governed by rules to achieve a task. In an SI system, each agent acts autonomously, reacts on dynamic inputs, and works collaboratively with other swarm members without a central control. The system as a whole is expected to exhibit global patterns and behaviors. It is the self-organization of systems for collective decentralized behavior and it follows a bottom-up approach.

The typical swarm intelligence system has the following properties [6]:

- It is composed of many individuals.
- The individuals are relatively homogeneous (i.e., they are either all identical or they belong to a few typologies).
- Action is stochastic choice based.
- The interactions among the individuals are based on simple behavioral rules that exploit only local information that the individuals exchange directly or via the environment.
- The overall behavior of the system results from the interactions of individuals with each other and with their environment, i.e., the group behavior self-organizes.

The characterizing property of a swarm intelligence system is its ability to act in a coordinated way without the presence of a coordinator or of an external controller. Based on the above properties, it is possible to design swarm intelligence system that are scalable, parallel, and fault tolerant.

## 17.4 SI Techniques

Swarm intelligence-based techniques are recent computational tools for systems optimization. They are population-based stochastic methods used in combinatorial optimization problems. The main characteristic of SI algorithms is that they automatically evolve by stimulating the social behavior of organisms. Researchers have developed various algorithms by modeling the behaviors of different swarm of animals. In other words, attempts to mimic the behavior of social insect swarms often lead to different implementations of swarm intelligence. More than 40 algorithms have been developed and more algorithms are appearing. Each algorithm has its own unique advantages. The most popular algorithms are the following [7–9]:

1. *Particle Swarm Optimization* (PSO): This type of algorithm is inspired by the flocking behavior of birds and the schooling behavior of fish in nature. PSO is a population-based algorithm for handling problems in which a point best represents a solution. It searches for an optimal solution in the computable search space. It is inspired by the intelligent, experience-sharing, social behavior of birds flocking to find the best food area, schools of fish, swarms of bees, or even human social behavior.
2. *Ant Colony Optimization* (ACO): This algorithm is based on the actions of natural ants. When an ant finds a source of food, it marks the path with chemical substances known as pheromones, which attract other ants to that path. ACO is a metaheuristic for solving combinatorial optimization problems. It draws inspiration from the behavior of ant colonies. While exploring their environment, real ants lay down pheromones directing each other to where resources are located.
3. *Swarm intelligence algorithms* (SIAs): In SIAs, a swarm is made up of multiple artificial agents. SIAs are essentially iterative-based stochastic search algorithms where heuristic information is shared. Instances of SIAs include bacterial foraging optimization (BFO), bee algorithm, fish schooling search, fireworks algorithm (FWA), and brain storm optimization (BSO).
4. *Bees Algorithm* (BA): This is inspired by the behavior of the honey bees. The hive sends out the scout bees which locate food source, return to the hive, and communicate with the other bees about the location of food source.
5. *Fireworks Algorithm* (FA): This is inspired by observing the firework explosion. In the FA, two explosion (search) processes are employed and mechanisms for keeping the diversity of sparks are also well designed.
6. *Gravitational Search Algorithm* (GSA): This stochastic search algorithm is a new type of approach based on Newton's concept of gravity: Every particle in the universe attracts every other particle with a force that is directly proportional to the product of their masses and inversely proportional to the square of the distance between them. The search agents are a collection of masses which interact with each other based on the Newtonian gravity. The masses correspond to good solutions of the problem.
7. *Bacterial Foraging Optimization Algorithm* (BFOA): This algorithm was inspired by the bacteria *E. Coli*. *E. Coli* bacteria forage in a way that it allows it to move towards or away from specific signals by perceiving chemical gradients in the area.

Besides these popular SI algorithms, others include evolutionary programming, genetic algorithm, firefly algorithms, artificial bee colony, fish swarm optimization, differential evolution, glowworm swarm optimization, and cuckoo search algorithm [10]. Figure 17.3 shows the hierarchical structure of SI Algorithms. These methods have been proposed to train artificial neural networks and other real-world problems

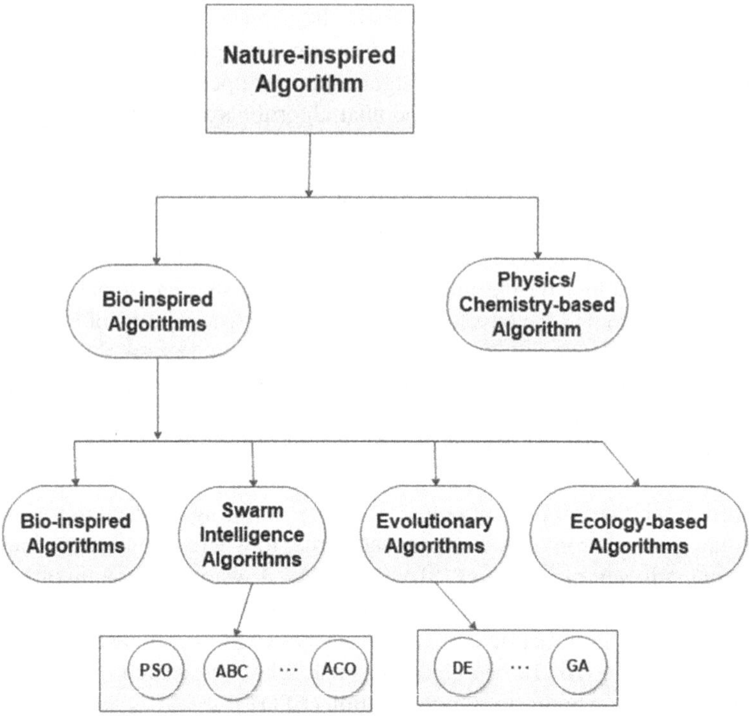


**Fig. 17.3** Hierarchical structure of SI algorithms

in science and engineering [7, 8]. These SI techniques may be regarded as another set of soft computing techniques.

## 17.5 Applications

Swarm intelligence-based algorithms have been applied in almost all areas of sciences and engineering. Although SI implementations in the real world are limited, SI techniques potentially find application in a wide spectrum of practical scenarios. SI is being applied in various settings to improve the management and control of large numbers of interacting entities such as communication, wireless sensor networks, and satellite constellations. The US military is investigating the use of SI in controlling unmanned vehicles. SI is used for solving problems in a variety of areas which include power system, fuzzy system, bio-medicine, sociological analysis, image processing, sound processing, signal processing, data analysis, process modeling, optimization, telecommunicate systems, robotics, military, process controlling, engineering, education, MANET, and forecasting. We will consider some of these applications.

- *Swarm Robots*: The application of swarm techniques to robots is called swarm robotics. Robots can walk, move, or operate under water. Swarm robotics may be regarded as the study of how a swarm of simple agents can be constructed to collectively accomplish tasks that are beyond the capability of a single one. It is a novel approach for coordinating a large number of robots [11]. A robot swarm is an autonomous entity that acts in a self-organized manner. It does not have centralized entity that is responsible for its coordination. Designing the robot-to-robot or robot-to-environment interactions that would result in the desired collective behavior is very challenging. It is possible to use swarm intelligence to control nanobots within the body for the purpose of killing cancer tumors. SI is well suited for multi-robot motion planning [12, 13]. A swarm of robots are shown in Fig. 17.4 [14].
- *Artificial Swarm Intelligence* (ASI): This is a means of amplifying the collective intelligence of networked human groups using control algorithms modeled after natural swarms. On one hand, we have natural swarm intelligence research, where biological systems are studied. On the other hand, we have artificial swarm intelligence, where human artifacts are studied. The ASI method, also known as swarm AI, has been shown to significantly amplify human intelligence. It has been used to enable groups of doctors to generate diagnoses with high accuracy. ASI addresses the limitations associated with group decision-making, amplifies the intelligence of human groups, and facilitates better decision-making process. The ASI interface is flexible [15].
- *Engineering*: SI techniques are useful tools for solving practical engineering problems. SI algorithms have been applied successfully to solve optimization problems in mechanical engineering [12]. SI has also been successfully applied to solve optimization problems in transportation engineering [16]. Swarm engineering approach uses a large number of simple robots whose interactions cause a global behavior.

**Fig. 17.4** Swarm robots [14]

- *Education*: In the educational system, swarm intelligence has been applied to curriculum planning, computer adaptive testing, intelligent assessment paper generation, and class/examinations scheduling [17]. Swarm intelligence can be adopted into a library system so that the system becomes intelligent. As long as the readers can access the Internet, they can conveniently obtain book information in the library that is closest to them [18].
- *Business:* Business systems that mimic swarm behavior have been used to solve optimization problems like freight logistics, routing in telecommunication networks, and planning workflow in distribution warehouses. The shifting dynamics of the field of swarm intelligence makes it critical for businesses to keep up with the changing market.
- *Human Resources:* When hiring, companies follow the practice of insects in hunting for food. Group recruiting is effective where competition is low and the labor market size is small. Using this mass-recruitment strategy, companies can hire the best talent even before other competitors emerge.
- *MANET*: As shown in Fig. 17.5, a mobile ad hoc network (MANET) is a self-organizing multihop wireless network where the structure of network changes dynamically. Routing and mobility are the most challenging problems in MANET. Routing in MANET is very difficult due to its dynamic topology and limited bandwidth. A new set of routing protocols based on swarm intelligence for MANETs has emerged. They are more capable of handling various routing problems such as scalability, maintainability, energy utilization, and high mobility [19, 20].

**Fig. 17.5** A mobile ad hoc network (MANET)

- *Medicine:* The ASI technology can be applied in medicine. This technology connects groups of human participants into real-time systems modeled after natural swarms and moderated by AI algorithms. SI-based technologies are very promising for use in diagnosis, enabling groups of medical professionals to amplify their insights and achieve diagnostic accuracies which outperform the accuracies of individual practitioners. Groups of networked radiologists can improve their diagnostic accuracy when reviewing chest X-rays for the presence of pneumonia. In ophthalmology, algorithms trained by machine learning can be as good as human experts in detecting a wide range of eye conditions [21]. There is an increased interest to overcome complex problems such as in cancer cure.
- *Cybersecurity:* In the digital era, transferring data is always vulnerable to attacks. Cybersecurity has become an indispensable component of computer systems with the objective of protecting valuable resource stored on systems from adversary attack. Applying SI in cybersecurity is based on the fact that machine learning approaches have been widely used to improve the efficiency of cyber defense systems. Typical SI-based algorithms that leverage in cyber domain include ant colony optimization, particle swarm optimization, artificial fish swarm, and artificial bee colony. These algorithms are implemented in various fields of cybersecurity. Scientists have proposed numerous approaches that have utilized SI to ensure integrity, confidentiality, and availability of information [22].
- *Wireless Sensor Network* (WSN): WSN has attracted a lot of attention due to its wide-ranged applications such as battlefield surveillance, machine failure diagnosis, biological detection, home security, smart spaces, and inventory tracking. A WSN consists of randomly distributed autonomous devices or nodes, with each sensor node having a CPU, battery supply, and a radio transceiver for communication. Sensors in a WSN are usually deployed in hostile environment such as desert or battlefield. Each sensor node is treated as ants for sensor nodes with basic capabilities to mimic the behavior of an ant. An SI-based energy saving greedy routing algorithm can be used to prolong the lifetime of a WSN [23].

In addition, SI can be used to solve the complex combinatorial optimization problems, nonlinear design system optimization, unmanned underwater vehicles, chemometrics, bioinformatics, biometric features selection and optimization, stock market price movement, face recognition, meeting scheduling, water resources management, routing in telecommunication networks, clustering ensemble, traffic and transportation, blockchains, HVAC (heating, ventilation, and air conditioning) systems, VLSI routing, anomaly detection systems, and data mining [24].

## 17.6 Benefits and Challenges

Swarm intelligence can be regarded as a mechanism which individuals can use to overcome some of their own cognitive limitations. Animals and humans make better decisions collectively. SI enables groups of entities to make collective decisions with significantly greater accuracy than the individuals could do by themselves. SI systems have the following advantages [25]: scalability, adaptability, flexibility, collective robustness, and individual simplicity. We can exploit the social collective behavior of swarms to solve complex real-life problems. Swarm intelligence boasts a number of attractive features such as adaptation, robustness, decentralized and self-organizing nature, which are suited for routing in modern communication networks [26].

The use of SI-based systems comes with a number of challenges. Predictability and controllability are hard to achieve in complex systems such as SI. For large and complex engineering problems, SI algorithms take a lot of computation time due to the stochastic nature of their searching approaches. Another disadvantage of swarm intelligence is its speed. Decisions where speed is a critical factor may show a simpler network of inter-individual interactions to minimize the number of connections and maximize response speed. Engineering design of systems based on the swarm intelligence paradigm can be challenging. The maintenance of a deployed swarm system is also challenging. A lot of work needs to be done before dependable robotic swarms can become an engineering reality [27].

## 17.7 Conclusion

Swarm intelligence is the collective behavior of decentralized, self-organized systems. It is inspired by social insects and has emerged in recent years as an innovative AI technique for solving complex problems. The concept refers to a kind of problem-solving ability that emerges in the interactions of simple information-processing units. The information-processing units that compose a swarm can be animals, insects, birds, human beings, robots, or stand-alone workstations [28]. Swarm intelligence is giving birth to collective intelligence, which has shown that groups can outperform individuals when making decisions, predictions, and forecasts.

The backbone of SI is built on two families of algorithms: ant colony optimization and particle swarm optimization. For more information, one should consult [13, 27, 29–53]. One should also consult the three journals exclusively devoted to swarm intelligence:

- *International Journal of Swarm Intelligence Research*
- *Swarm and Evolutionary Computation*
- *Swarm Intelligence*

## References

1. M.N.O. Sadiku, M. Tembely, S.M. Musa, Swarm intelligence. Int. J.. Adv. Res. Comp. Sci. Soft. Eng. **8**(5), 100–102 (2018)

2. G. Di Caro, An introduction to swarm intelligence issues, http://staff.washington.edu/paymana/swarm/dicaro_lecture1.pdf
3. J.H. Rolling, Swarm intelligence and collaboration. Art Educ. **69**(5), 4–6 (2016)
4. Swarm intelligence, Wikipedia, the free encyclopedia, https://en.wikipedia.org/wiki/Swarm_intelligence
5. A. Íñiguez, The five principles of swarm intelligence (2016), https://www.linkedin.com/pulse/five-principles-swarm-intelligence-alfonso-%C3%AD%C3%B1iguez
6. M. Dorigo, M. Birattari, Swarm intelligence. Scholarpedia **2**(9) (2007)
7. M. Conforth, Y. Meng, Reinforcement learning using swarm intelligence-trained neural networks. J. Exp. Theor. Artific. Intell. **22**(3), 197–218 (2010)
8. M.J.R. Mish, Swarm intelligence techniques and its applications in water resources management. ISH J. Hydrau. Eng. **15**(Suppl 1), 151–169 (2009)
9. 5 Examples of swarm intelligence, https://www.bestcomputersciencedegrees.com/lists/5-examples-of-swarm-intelligence/
10. M.N.A. Wahab, S. Nefti-Meziani, A. Atyabi, A comprehensive review of swarm optimization algorithms. PLoS One **10**(5), e0122827 (2015)
11. G.G. Rigatos, Multi-robot motion planning using swarm intelligence. Int. J. Adv. Robot. Syst. **5**(2), 139–144 (2008)
12. B. Yu, G. Chen, Swarm intelligence in mechanical engineering. Adv. Mech. Eng. **8**(12), 1–3 (2016)
13. E. Sahin, W. M. Spears (eds.), *Swarm Robotics* (Springer, New York, 2005)
14. Swarm intelligence, http://www.techferry.com/articles/swarm-intelligence.html
15. L. Metcalf, D.A. Askay, L.B. Rosenberg, Keeping humans in the loop: Pooling knowledge through artificial swarm intelligence to improve business decision making. Calif. Manag. Rev. **61**(4), 84–109 (2019)
16. Swarm intelligence in transportation engineering. Adv. Mech. Eng. **11**(3), 1–3 (2019)
17. L.S. Chen, Applying swarm intelligence to a library system. Libr. Collect. Acquis. Tech. Serv. **34**(1), 1–10 (2010)
18. C.C. Ioannou, Swarm intelligence in fish? The difficulty in demonstrating distributed and self-organised collective intelligence in (some) animal groups. Behav. Process. **141**, 141–151 (2017)
19. N. Srivastava, P. Raghav, A review on swarm intelligence based routing algorithms in mobile adhoc network, in *Proceeding of the 8th International Conference on Computing, Communication and Networking Technologies*, Delhi (2017)
20. A. Giagkos, M.S. Wilson, Swarm intelligence to wireless ad hoc networks: Adaptive honeybee foraging during communication sessions. Adapt. Behav. **21**(6), 501–515 (2013)
21. L. Rosenberg et al, Artificial swarm intelligence employed to amplify diagnostic accuracy in radiology, in *Proceedings of IEEE 9th Annual Information Technology, Electronics and Mobile Communication Conference (IEMCON)* (2018)
22. T.C. Truong, T.P. Huynh, I. Zelinka, Applications of swarm intelligence algorithms countering the cyber threats, in *Proceeding of GECCO'20 Companion, Cancún, Mexico* (2020), pp. 1476–1485
23. W. Rui, et al., Swarm intelligence for the self-organization of wireless sensor network, in *Proceedings of IEEE Congress on Evolutionary Computation*, Vancouver, BC, Canada (2006), pp. 838–842
24. A. Saggu, P. Yadav, M. Roopak, Applications of swarm intelligence. Int. J. Comput. Sci. Mob. Comput. **2**(5), 353–359 (2013)
25. H. Ahmed, J. Glasgow, Swarm intelligence: concepts, models and applications, https://pdfs.semanticscholar.org/116b/67cf2ad2c948533e6890a9fccc5543dded89.pdf
26. Y. Lv, Y. Zhu, Routing algorithm based on swarm intelligence, in *Proceedings of the 10th World Congress on Intelligent Control and Automation*, Beijing, China (2012)
27. J. Kennedy, Smarm Intelligence, in *Handbook of Nature-Inspired and Innovation Computing*, ed. by A. Y. Zomaya, (Springer, New York, 2006), pp. 187–219

28. M. Schranz et al., Swarm intelligence and cyber-physical systems: concepts, challenges and future trends. Swarm Evol.Comput. **60**, 100762 (2021)
29. A. Abraham, C. Grosan, V. Ramos (eds.), *Swarm Intelligence in Data Mining* (Springer, 2006)
30. X.S. Yang, *Nature-Inspired Computation and Swarm Intelligence: Algorithms, Theory and Applications* (Academic Press, 2020)
31. A. Nayyar, D. N. Le, N. G. Nguyen (eds.), *Advances in Swarm Intelligence for Optimizing Problems in Computer Science* (Chapman and Hall/CRC, 2018)
32. J. Kennedy, R.C. Eberhart, Y. Shi, *Swarm Intelligence* (Morgan Kaufmann, San Francisco, CA, 2001)
33. E. Bonabeau, M. Dorigo, G. Theraulaz, *Swarm Intelligence: From Natural to Artificial System* (Oxford University Press, New York, 1999)
34. S. Bhattacharyya, P. Dutta (eds.), *Handbook of Research on Swarm Intelligence in Engineering* (Engineering Science Reference, 2015)
35. A.E. Hassanien, E. Emary, *Swarm Intelligence: Principles, Advances, and Applications* (CRC Press, Boca Raton, FL, 2017)
36. Y. Shi (ed.), *Emerging Research on Swarm Intelligence and Algorithm Optimization* (Information Science Reference, 2015)
37. Y. Tan, Y. Shi, T. K. Chen (eds.), *Advances in Swarm Intelligence* (Springer, Berlin, 2010)
38. B. K. Panigrahi, Y. Shi, M. H. Lim (eds.), *Handbook of Swarm Intelligence: Concepts, Principles and Applications* (Springer, 2011)
39. S. Dehuri, A.K. Jagadev, M. Panda, *Multi-Objective Swarm Intelligence: Theoretical Advances and Applications* (Springer, 2020)
40. Y. Tan, Y. Shi, M. Tuba, *Buy Advances in Swarm Intelligence* (Springer, 2020)
41. X.S. Yang et al., *Buy Swarm Intelligence and Bio-Inspired Computation: Theory and Applications* (Elsevier Science, 2020)
42. Y. Shi, *Innovations and Developments of Swarm Intelligence Applications*. IGI Global, 2012
43. Y. Shi, *Critical Developments and Applications of Swarm Intelligence*. IGI Global, 2018
44. F. Chan, M. Tiwari (eds.), *Swarm Intelligence: Focus on Ant and Particle Swarm Optimization* (InTech, 2007)
45. J.B. Waldner, *Nanocomputers and Swarm Intelligence* (Wiley, 2013)
46. C. P. Lim, L. C. Jain, S. Dehuri (eds.), *Innovations in Swarm Intelligence* (Springer, 2009)
47. N. Nedjah, L. D. S. Coelho, L. M. Mourelle (eds.), *Multi-Objective Swarm Intelligent Systems: Theory & Experiences* (Springer, 2009)
48. B. Muller, *Theory and New Applications of Swarm Intelligence* (Scitus Academics, 2006)
49. A. Slowik (ed.), *Swarm Intelligence Algorithms: A Tutorial* (CRC Press, Boca Raton, FL, 2020)
50. L. P. Walters (ed.), *Applications of Swarm Intelligence* (Nova Science Publishers, 2010)
51. R. Eberhart, Y. Shi, J. Kennedy, *Swarm Intelligence* (Morgan Kaufmann, 2001)
52. A.P. Engelbrecht, *Fundamentals of Computational Swarm Intelligence* (Wiley, 2006)
53. C. Blum, D. Merkle (eds.), *Swarm Intelligence: Introduction and Applications* (Springer, 2008)

# Chapter 18
# Ambient Intelligence

**Contents**



> Ambient intelligence is the convergence of ubiquitous computing, ubiquitous communication, and interfaces adapting to the user
>
> —Gupta

## 18.1 Introduction

In recent years, there has been significant advances in three promising technology areas: virtual environments, mobile communication, and sensors. The merging of these areas allows the emergence of a new vision: the Ambient Intelligence (AmI), in which people are empowered through a digital environment that is sensitive, adaptive, and responsive to human needs [1]. Ambient Intelligence is an environment that intelligently supports people inhabiting in it. It proposes a new way to interact between humans and technology.

Ambient intelligence is regarded as one of the emerging technologies that work with artificial intelligence techniques and wireless sensor networks to bring intelligence to our everyday environments and make those environments sensitive to humans. Its goal is to enhance the way people interact with their environment to promote safety and to enrich their lives. People will be surrounded by intelligent objects that can sense the context and respond according to their desire.



M. N. O. Sadiku, S. M. Musa, *A Primer on Multiple Intelligences*,
https://doi.org/10.1007/978-3-030-77584-1_18

Ambient intelligence (AmI), also known as intelligent environments (IEs), refers to electronic environments that are sensitive and responsive to the presence of people (patients, doctors, nurses, and informal caregivers). While "intelligence" refers to the ability to learn and apply knowledge in new situations, "ambient" describes a physical space (hospitals, clinics, and assisted living facilities, etc.) and its internal and external surroundings [2]. The ambient devices offer interfaces that are easy to use and do not require prior learning. Ambient intelligence may be regarded as a data-intensive environment that senses changes in state and responds appropriately to correct, act, or alert decision-makers. It is any digitized working environment that is designed with embedded technology and AI to assist people. The environment is unobtrusive, interconnected, embedded, and intelligent. The environment is sensitive to the needs of its inhabitants and capable of anticipating their needs and behavior.

This chapter provides an introduction on the use of ambient intelligence. It begins by explaining the concept of ambient intelligence, its characteristics, and its enabling technologies. It presents some applications of AmI. It considers AmI in the global context. It addresses the benefits and challenges of AmI. The last section concludes with comments.

## 18.2 Concept of Ambient Intelligence

Today, we are entering the age of "embedded-everywhere" in which objects around us become intelligent interacting with each other and with people through wireless sensors and actuators. This culminates in the idea of Ambient intelligence (AmI), which is the vision that technology will become invisible, embedded in our natural surroundings, and support the people inhabiting them. It refers to an intelligent, embedded, digital environment that is sensitive and responsive to the presence of people. It is intelligence gathered from a wide variety of sources in our personal environment, such as sensors, wearables, and smart devices.

Ambient Intelligence (AmI) is a new paradigm in information technology. It is simply intelligence embedded in the environment. It may be perceived as an evolution of smart environment systems. It may also be regarded as a computing paradigm wherein conventional input and output media no longer exist. AmI can be implemented by integrating sensors, cameras, microphones, and other devices which can generate signals. This fundamentally changes the human–machine interaction: the traditional PC input and output media such as mice, keyboards, and screens are disappearing. Instead, sensors and processors are integrated into conventional objects that harmonize with people in their living environment [3]. As shown in Fig. 18.1, AmI environments can be diverse such as homes, stores, offices, schools, restaurants, hospitals, vehicles, factories, sports facilities, and churches. Figure 18.2 shows what ambient intelligence is all about.

The term "ambient intelligence" was coined by the European Commission in 2001. It was developed in the late 1990s in a series of internal workshops conducted

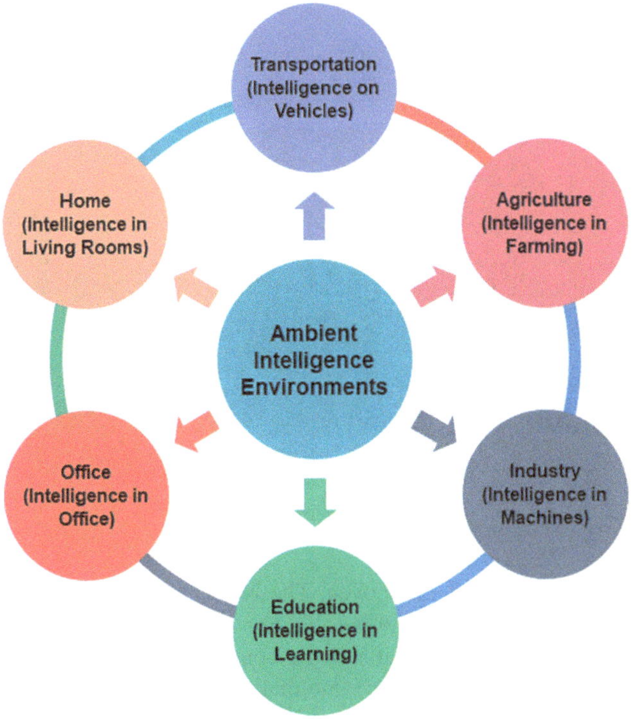


**Fig. 18.1** Typical ambient intelligence environments

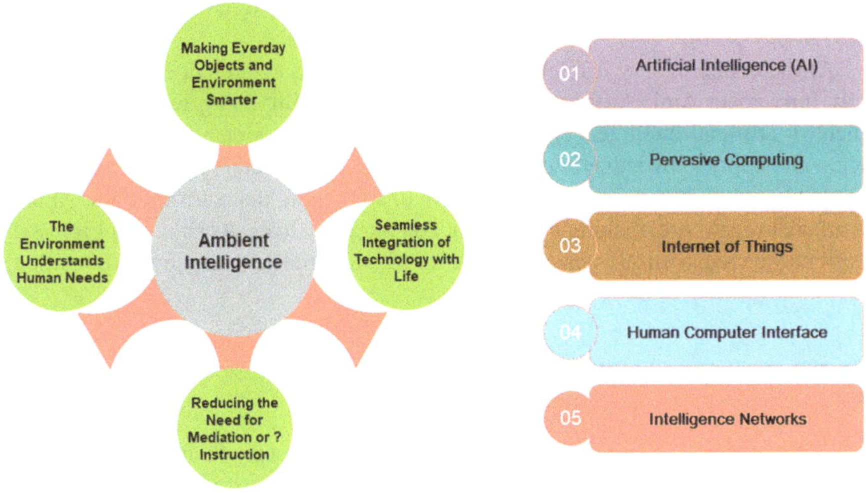


**Fig. 18.2** What is ambient intelligence?

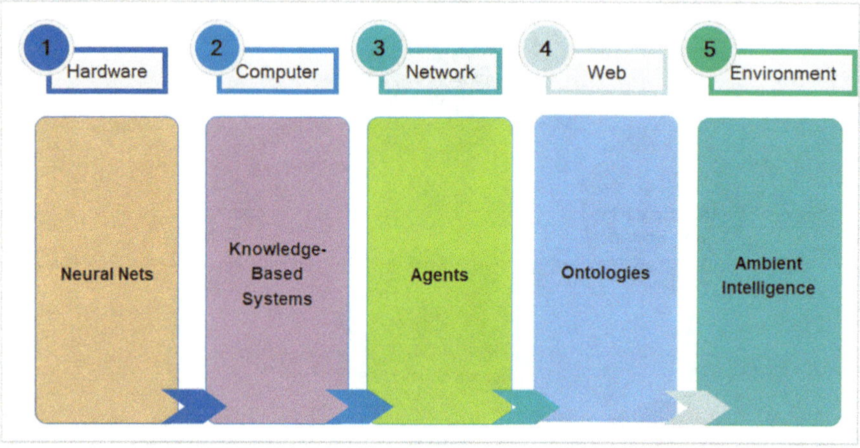


**Fig. 18.3** The evolution of artificial intelligence

by Philips (headquartered in Eindhoven, The Netherlands). The workshops were aimed at understanding the need to put the user in the center of product development. It soon became apparent that a single company could not possibly turn the broad concept of ambient intelligence into reality. Although it was originally envisioned for consumer electronics, ambient intelligence has grown into other areas such as healthcare, public, and productivity domains. AmI will be one of the key elements of the fourth industrial revolution. In 2001, AmI was taken up by The European Commission's Information Society Technologies Advisory Group (ISTAG).

Artificial intelligence is related to ambient intelligence, but it does not conceptually fully cover AmI. Ambient intelligence is developing along with the development of artificial intelligence. Intelligence refers to man's or machine's ability to learn and apply knowledge in new situations. While "artificial" refers to something made by human beings, "ambience" refers to something that surrounds us. AI and AmI are everywhere, making a big impact on our daily lives. As shown in Fig. 18.3, AmI may be regarded as a new challenge for AI and is the next step in AI's evolution.

## 18.3 Ambient Intelligence Characteristics

Ambient intelligence (AmI) is a multidisciplinary approach which aims to enhance the way environments and humans interact with each other. It is essentially a technology that combines Internet of things (IoT) and artificial intelligence (AI). It is intrinsically connected with AI. Practically, AmI may be roughly regarded as the opposite of virtual reality. While virtual reality puts people inside a computer-generated world, AmI puts the computer inside the world to help people.

The Information Society Technologies Advisory Group (ISTAG) identified a series of necessary characteristics that will permit wide societal acceptance of AmI. AmI should have the following characteristics [4]:

- AmI should facilitate human contact.
- AmI should be orientated towards community and cultural enhancement.
- AmI should help to build knowledge and skills for work, better quality of work, citizenship, and consumer choice.
- AmI should inspire trust and confidence.
- AmI should be consistent with long-term sustainability—personal, societal, and environmental—and with lifelong learning.
- AmI should be made easy to live with and controllable by ordinary people.

An AmI system has following characteristics [5, 6]:

- *Context aware*: It can recognize the situational context of subject.
- *Personalized:* It is personalized and tailored to the needs of each individual.
- *Anticipatory:* It can anticipate the needs or desires of an individual without the conscious mediation of the individual.
- *Adaptive*: It adapts to the changing needs of individuals.
- *Ubiquity:* It is embedded and is integrated into our everyday environments.
- *Transparency:* It recedes into the background of our daily life in an unobtrusive way.
- *Intelligent:* It uses intelligence gathered from a wide variety of sources in our personal environment.

## 18.4 Enabling Technologies

In order for AmI to become a reality, a number of key technologies are required. Ambient intelligence is a technology-enriched environment which involves extensive and invisible integration of technologies in our everyday lives: smart phones, tablets, wireless sensor network (Wi-Fi, Bluetooth, NFC, RFID, etc.), Internet (Facebook, WhatsApp, Twitter, You Tube, Blogs, Cloud Computing, etc.), data acquisition systems, robotic systems, communication technology, wearable devices, decision support, analytics, machine actuators, machine learning, AI, human–computer interaction (HCI), wireless sensor networks (WSNs), and other capabilities [4]. Its five components are: (1) IoT sensors, (2) human–computer interactions, (3) pervasive computing, (4) AI frameworks, and (5) invisible intelligent network. Figure 18.4 shows the confluence of different areas into ambient intelligence.

Wireless sensor networks (WSNs) are used for gathering the information needed by AmI environments. Typical examples of wireless technologies are radio frequency identification tags (RFID), Wi-Fi, ZigBee, or Bluetooth [7]. Standards such as ZigBee, ZWave, and Bluetooth provide wireless communication interfaces between the devices. The principal device in WSNs is the network node. This device

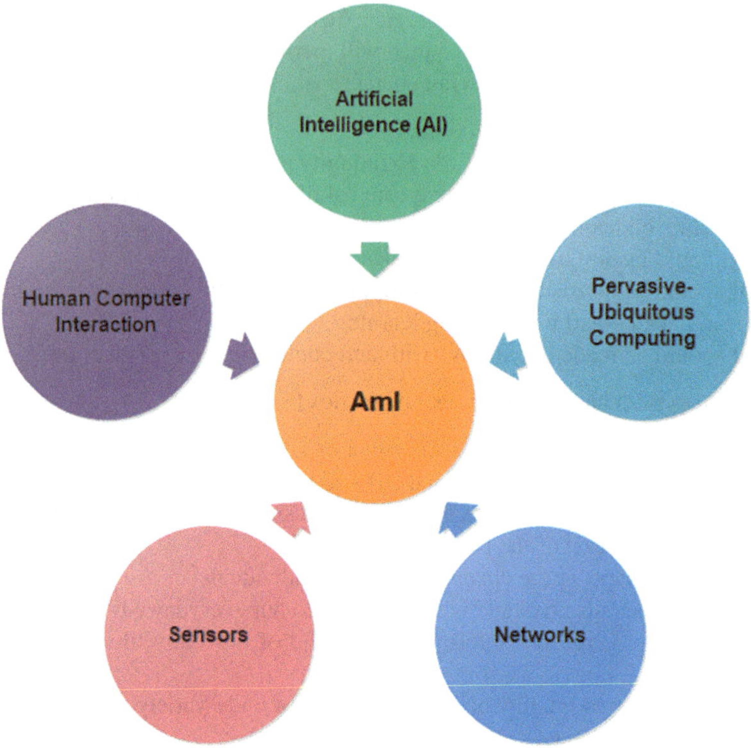


**Fig. 18.4** Areas related to ambient intelligence

has RFID for the transmission and the reception of information. Sensor data is collected from several sources and later analyzed to produce accurate information.

## 18.5 Applications

Ambient intelligence has several potential applications in diverse areas such as healthcare, smart home, office, transport, industry, entertainment, tourism, recommender systems, safety systems, ehealth, supported living, assisting medical diagnosis, e-learning, education, culture, entertainment, medicine, robotics, smart classrooms, and smart cities [8]. Some of these are explained as follows.

- *Healthcare:* Ambient intelligence healthcare systems are increasingly being deployed in hospitals, clinics, and nursing homes. Common healthcare applications include monitoring the status of patients with chronic diseases, assistive care, persuasive services for encouraging healthier lifestyles, rehabilitation, and emergency services. It is also applicable to patient monitoring and neuroscience research, which require large biomedical data sets. Figure 18.5 illustrates some

*AmI health services*. Ambient assisted service anywhere and anytime is an AI-based technology that provides safe, high-quality, and independent lives for the frail and elderly to ensure their health and happiness. Assisted living systems must monitor, control, and interact with a number of different components of the living environment, without interfering with the lives of those being monitored. Elderly and disabled people would benefit much from AmI if it is accessible. Individuals with disabilities can live independently using home automation that provides constant monitoring [9].

- *Smart hospital:* In traditional hospitals, healthcare practitioners must navigate the hospital to gather information about patients and control their status. A smart hospital is a smart environment that employs ambient intelligence technologies in order to augment a typical hospital room with smart features that assist both patients and medical staff. This environment allows the patient to control the environment and interact with the hospital facilities [10].
- *Smart Home*: A great deal of the vision of AmI is targeted at smart homes. The concept of a smart home refers to a future home equipped with technical devices that reduce burdens and provide better quality of life. It is a home equipped with sensors and activators to monitor activities and movement. The idea extends to hospitals, public transport, factories, and other environments. Ambient intelligence coupled with information and communication technologies (ICTs) is increasingly used in the home environment to facilitate independent living for older elder citizens. AmI solutions at home will cover four areas: (a) Home automation (or domotics), (b) Communication and socialization, (c) Rest, refreshing,

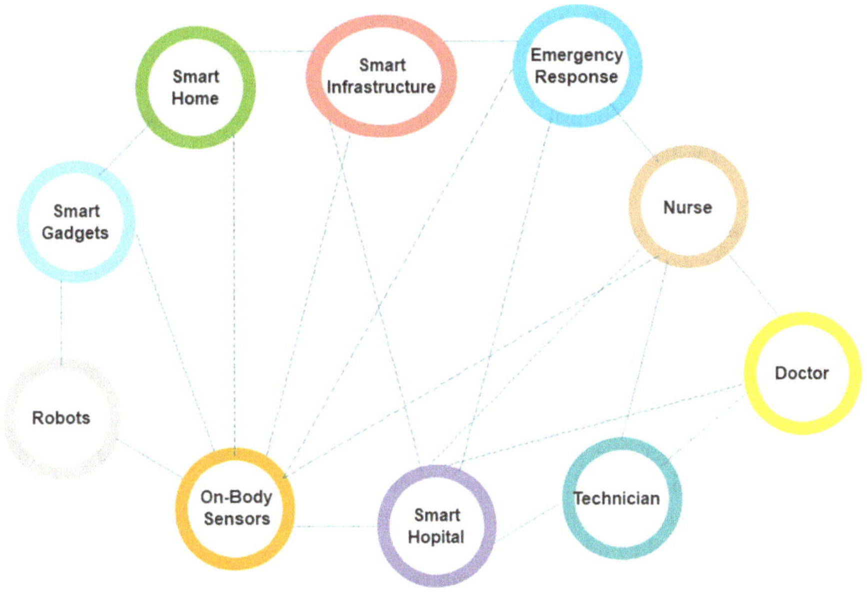


**Fig. 18.5** Interconnected world of AmI health service

and sport, (d) Household work and learning. For example, the Microsoft Corporation Domestic Home in Redmond, Washington showcases AmI-based smart appliance technologies for the home [11].

- *e-Learning:* Traditionally, e-Learning seldom tailors itself to each student. An AmI-assisted e-learning platform provides an adaptive and smart learning environment that can respond to the needs of each learner. With the combination of the appropriate AmI techniques, mobile computing, and sensing technologies, it is possible to determine the individual's actual progress and response to the learning materials, possibly delivered anytime and anywhere on mobile devices. AmI techniques work with the learners' characteristics and learning needs to come up with a more effective or personalized e-learning platform [12]. The learning material (e.g., books or manuals) could be an integral part of the ambient learning environment. Learners have access to their documents from anywhere at any time.
- *Education:* Academic institutions may use technology to track student's progress on their tasks and frequency of attendance to classes. AmI can improve the learning experience for students. The intelligent classroom employs some devices to infer speaker intent. From the inferred intent, the room can control light settings, play videos, and display slides [13].
- *Transportation:* We spend a significant part of our life traveling. Public transportation can benefit from AmI technologies including GPS-based spatial location, vehicle identification, and image processing. This can allow a system to assist the driver more effectively when the help is most needed. Emergency services like fire brigades can improve using AmI technologies [14].
- *Logistics:* Companies have been using sensors for tracking and tracing comparatively early. Sensors offer the opportunity to design a smart warehouse that identifies the incoming and outgoing goods automatically and directly orders new goods if necessary. RFID (radio frequency identification) technology is being increasingly used by industry for tracking inventory. These tags can be small and do not require a battery.

Some of these applications of ambient intelligence are depicted in Fig. 18.6. Other applications of ambient intelligence include smart cities, emergency services, mechatronics systems, hospitals, and elderly people's homes.

## 18.6 Global Ambient Intelligence

As mentioned earlier, the concept "ambient intelligence" was introduced by the European Commission's Information Society Technologies Advisory Group (ISTAG) in 2001. Ambient intelligence is a new paradigm in information technology compliant with the international Integrating Healthcare Enterprise (IHE) board. AI is considered the backbone of the ambient intelligence technology. AmI seeks to effect a significant change in the way people live their daily life. In this section, we

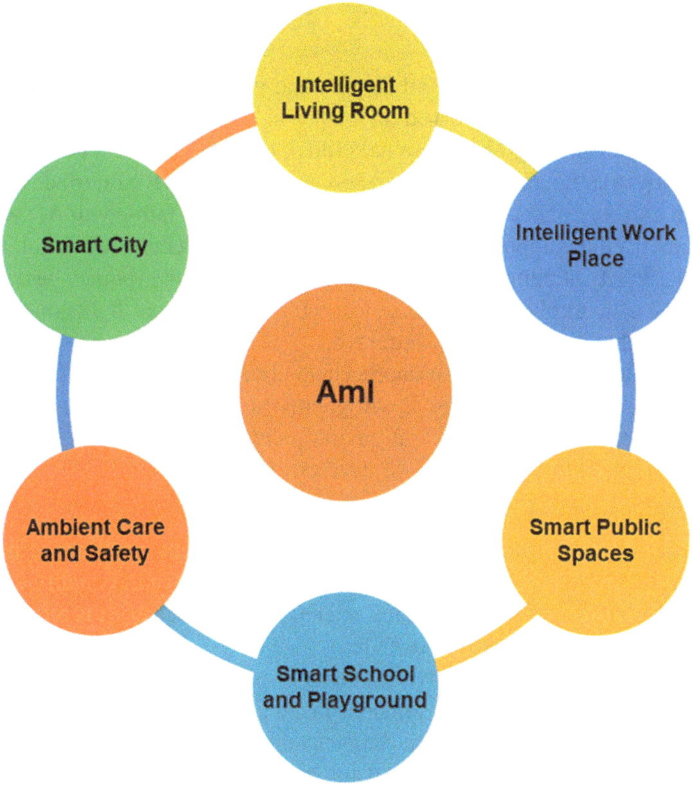


**Fig. 18.6** Some applications of ambient intelligence

examine how cultural, social, and environmental factors determine the evolution of ambient intelligence in different parts of the world. Although Western and Eastern cultures are different, AmI in both cultures has similar approach on assuring the safety of users. However, there are some differences which allow us to explore the idea of localized ambient and aesthetic intelligence [15].

The segments of population that will benefit most from AmI systems are the elderly and people with disabilities. It will improve important aspects of their life, especially healthcare. There is an ever growing need to provide effective and affordable care and support to the disabled and elderly. Today, the number of Europeans over 60 years represents more than 25% of the population. In the United States, it is expected that in 2020 people over 60 will represent about 1 of 6 citizens.

The AmI vision has been used by Philips Research laboratories to establish new and promising collaborations with other strong players in the field. Personalization and context-awareness have been some of the objectives of AmI technologies developed at Philips, which has established itself as the intellectual leader in the field of ambient intelligence [16].

## 18.7 Benefits

Ambient intelligence is the potential for vast and varied applications, bringing with it both promise and peril. It is based on the premise that if environment can be personalized for the current needs of an individual, then the lives can be more productive and enjoyable. It can provide a wireless and more intuitive interface.Since AmI and AI are closely connected, AmI is sharing its fate and fame with AI. AmI technologies are used to support aging-in-place and they can contribute to an increased safety and security at home. Ambient intelligence can greatly reduce the occurrence of accidents in workplaces such as construction sites. It can fight loneliness and work in ways that are largely non-intrusive.

Ambient intelligence opens up unprecedented new perspectives. It can give us a better understanding of ourselves. It will greatly modify interaction with respect to current graphical environments. AmI environments will be very interaction intensive and will integrate a wide range of interactive devices.

Other benefits include the following:

- *Technology Integration*: The main objective of ambient intelligence is to integrate technology into every aspect of our lives to improve the quality of our life. Ambient intelligence serves humans, society, and the planet. It is a fantastic tool for improving and simplifying our life.
- *Monitoring:* AmI for healthcare monitoring and personalized healthcare is a promising solution to provide efficient medical services, which could substantially reduce the healthcare cost. Ambient intelligence enhances the quality of healthcare monitoring systems. AmI for health monitoring may be an effective solution to control elderly, disabled, and people with chronic illness anytime, anywhere and in an unobtrusive way with low cost.
- *Smart Environment:* Ambient intelligence-based systems can be developed to create smart environments for user and environmental monitoring. Ambient intelligence is capable of creating environments that are smart and proactive. These days, things are becoming smart: smart cities, smart hospital, smart environments, smart technologies, smart data, smart phones, smart rooms, smart homes, smart building, smart mobility, smart airports, smart nations, etc.
- *Productivity:* Ambient intelligent environment enhances the productivity and decision-making process of managers who work in complex and dynamic environments where critical decisions are constantly made. Digital assistance through AmI can help to free healthcare practitioners so that they can have time to take care of patients.
- *Elderly Care*: Elderly citizens can benefit from the AmI concept. AmI technology is a tool that can provide cost-effective ways to support the elderly individuals by monitoring them in their homes. The elderly can take advantage of the use of AmI systems because they can focus on older adults' real needs. They are given an opportunity to live more independently and for longer time in their home rather than in healthcare facilities. Independent living of elderly can be supported with mobile-centric ambient intelligence services.

## 18.8 Challenges

Ambient intelligence requires new software architectures that support large, distributed, and heterogeneous systems. For this reason, making AmI real is difficult task, especially when considering public spaces in urban environments. It is beset by problems such as lack of standardization, sustainability, and cost inefficiencies. The process of implementing an ambient system is a great challenge. Some other challenges include the following [17–19].

- *Trust and Confidence:* AmI needs to gain the confidence and trust of the public before it can be widely adopted. Individuals should be able to trust the intelligent world that surrounds them and through which they move. We expect privacy and confidentiality from those to whom we disclose our personal information.
- *Security and Privacy:* Ambient intelligence also gives rise to security and privacy concerns due the collection of large amounts of personal data involved. The success of ambient intelligence will be dictated by on how secure it can be made. Security issues are related to the health status of the patient, who may be frail and afraid that they are more vulnerable in cases of emergency. Privacy is also a major concern as ambient intelligence becomes more prevalent and is able to uniquely track people. AmI in the home environments is attended with concerns about privacy, sensitivity of data, and ethics.
- *Safety:* We are gradually moving into a new era where artificial intelligent agents become intertwined with our daily living. There is no guarantee that these agents can cooperate appropriately and safely.
- *Ethical Concern:* AmI has some serious ethical challenges. It is important for us to consider the ethical implications of using "hidden" intelligence to collect, process, analyze, and share personal data. A number of questions relating to home monitoring technologies have been raised. These include questions on how much surveillance is helpful, when does technology start infringing on personal dignity, and whether home automation can improve human condition without frustrating the user.
- *Abstract Concept*: The AmI vision is still abstract. It is not yet useful for funding decisions, research project definition, and business plan development. Its social benefits cannot be realized unless some requirements regarding mixed-technology design have been met.

These challenges must be addressed in order to develop flexible and evolvable AmI applications. AmI still needs to gain the trust and confidence of the users.

## 18.9 Conclusion

Ambient intelligence refers to a system and an information technology that can adapt to human activities in living environment. It describes an environment in which the technology is embedded and hidden in the background and interacts with people and objects. It is an emerging field that aims to bring intelligence to our everyday environments and make those environments sensitive to the inhabitants.

An ambient intelligence (AmI) revolution is already taking place. By transforming our ways of thinking and our interactions with the environment, the ambient intelligence opens up unprecedented new perspectives. The interest in ambient intelligence has increased drastically and worldwide due to the widespread use of portable devices. Its success depends heavily on the progress of ICT and AI. Students in engineering and computer science should be trained for the ambient intelligence challenge. More information about ambient intelligence can be found in the books in [1, 20–40] and the following closely related journals:

- *International Journal of Ambient Computing and Intelligence*
- *AI Communications*
- *Journal of Ambient Intelligence and Humanized Computing*
- *Journal of Ambient Intelligence and Smart Environment*
- *Intelligence*

## References

1. J. M. Corchado, D. I. Tapia, J. Bravo (eds.), *3rd Symposium of Ubiquitous Computing and Ambient Intelligence 2008* (Springer, Berlin, 2009), pp. 76–85
2. M.N.O. Sadiku, S.M. Musa, A. Ajayi-Majebi, Ambient intelligence: a primer. Int. J. Trend Res. Develop. **7**(5) (2020)
3. J. Esch, A survey on ambient intelligence in healthcare. Proc. IEEE **101**(12), 2467–2469 (2013)
4. Ambient intelligence, Wikipedia, the free encyclopedia, https://en.wikipedia.org/wiki/Ambient_intelligence
5. G. Acampora et al., A survey on ambient intelligence in healthcare. Proc. IEEE **101**(12), 2470–2494 (2013)
6. Ambient intelligence for health and cognitive enhancement, https://aaai.org/Press/Reports/Symposia/Spring/ss-15-01.php
7. A.A.M. Salih, A. Abraham, A review of ambient intelligence assisted healthcare monitoring. Int. J.Comput. Inform. Syst. Indust. Manag. Appl. **5**, 741–750 (2013)
8. J.C. Augusto, Ambient intelligence: Opportunities and consequences of its use in smart classrooms. Innov. Teach. Learn. Inform. Comput. Sci. **8**(2), 53–63 (2009)
9. M.N.O. Sadiku, S.M. Musa, A. Ajayi-Majebi, Ambient intelligence in healthcare. Int. J. Trend Res. Develop. **7**(5) (2020)
10. S. Kartakis, Enhancing health care delivery through ambient intelligence applications. Sensors **12**, 11435–11450 (2012)
11. F. Sadri, Ambient intelligence: a survey. ACM Comput. Surv. **43**(4) (2011)
12. M. Gams et al., Artificial intelligence and ambient intelligence. J. Ambie. Intell. Smart Environ. **11**(1), 71–86 (2019)

13. D.J. Cook et al., Ambient intelligence: Technologies, applications, and opportunities. Perv. Mob. Comput. **5**(4), 277–298 (2009)
14. M. Gams et al., Artificial intelligence and ambient intelligence. J. Amb. Intell. Smart Environ. **11**, 71–86 (2019)
15. C.L. Kaiyinga, D.A. Plewea, C. Röckerb, The ambience of ambient intelligence: An Asian approach to ambient systems? Proc. Manufactur. **3**, 2155–2161 (2015)
16. B. de Ruyter, E. Aarts, Ambient intelligence: visualizing the future, in *Proceedings of the Working Conference on Advanced Visual Interfaces*, Gallipoli, Italy (2004), pp. 203–208
17. J. van Hoof et al., Ageing-in-place with the use of ambient intelligence technology: Perspectives of older users. Int. J. Med. Inform. **80**(5), 310–331 (2011)
18. M. Lindwer et al., Ambient intelligence visions and achievements: Linking abstract ideas to real-world concepts, in *Proceedings of the Design, Automation and Test in Europe Conference and Exhibition* (2003)
19. E. Kosta, Mobile-centric ambient intelligence in health- and homecare—anticipating ethical and legal challenges. Sci. Eng. Ethics **16**(2), 303–323 (2010)
20. J. C. Augusto et al. (eds.), *Handbook of Ambient Assisted Living; Technology for Healthcare, Rehabilitation and Well-being* (IOS Press, 2012)
21. G.L. Foresti, T. Ellis, *Ambient Intelligence: A Novel Paradigm* (Springer Science, 2005)
22. E.H.L. Aarts, J.L. Encarnação, *True Visions: The Emergence of Ambient Intelligence* (Springer, 2006)
23. A. Salih, A. Abraham, *Ambient Intelligence Assisted Healthcare Monitoring* (LAP LAMBERT Academic Publishing, 2016)
24. R. Wichert et al., *Ambient Intelligence* (Springer, 2010)
25. K. Curran (ed.), *Recent Advances in Ambient Intelligence and Context-Aware Computing* (Information Science Reference, 2015)
26. J. Arends et al., *Recent Advances in Ambient Assisted Living - Bridging Assistive Technologies, E-Health and Personalized Health Care* (IOS Press, 2015)
27. A. Aztiria, J. C. Augusto, A. Orlandini (eds.), *State of the Art in AI Applied to Ambient Intelligence* (IOS Press, 2018)
28. H. Nakashima, H. Aghaja, J.C. Augusto, *Handbook of Ambient Intelligence and Smart Environments* (Springer, 2010)
29. S. Mukherjee et al., *AmIware Hardware Technology Drivers of Ambient Intelligence* (The Netherlands, Springer, Dordrecht, 2006)
30. Y. Cai, J. Abascal (eds.), *Ambient Intelligence in Everyday Life* (Springer, 2006)
31. G. Riva et al. (eds.), *Ambient Intelligence: The Evolution of Technology, Communication and Cognition Towards the Future of Human-Computer Interaction* (IOS Press, Amsterdam, 2005)
32. N.M. Garcia, J.J.P.C. Rodrigues, *Ambient Assisted Living* (CRC Press, Boca Raton, FL, 2015)
33. T. Bosse et al. (eds.), *Human Aspects in Ambient Intelligence: Contemporary Challenges and Solutions* (Atlantis Press, 2013)
34. T. Bosse (ed.), *Agents and Ambient Intelligence; Achievements and Challenges in the Intersection of Agent Technology and Ambient Intelligence* (IOS Press, 2012)
35. K. Curran (ed.), *Innovative Applications of Ambient Intelligence; Advances in Smart Systems* (Information Science Reference, 2012)
36. N. Y. Chong, F. Mastrogiovanni (eds.), *Handbook of Research on Ambient Intelligence and Smart Environments; Trends and Perspective* (Information Science Reference, 2011)
37. J.V. De Ven, *Supporting Communication in Spatially Distributed Groups: Privacy as a Service for Ambient Intelligence* (IOS Press, 2016)
38. E. Aarts, S. Marzano (eds.), *The New Everyday: Views on Ambient Intelligence* (010 Publishers, Rotterdam, 2003)
39. D. Wright et al. (eds.), *Safeguards in a World of Ambient Intelligence* (Springer, 2008)
40. Z. Mahmood (ed.), *Guide to Ambient Intelligence in the IoT Environment Principles, Technologies and Applications* (Springer, 2019)

# Chapter 19
# Musical Intelligence

**Contents**



Without music, life would be a mistake.

—Friedrich Nietzsche

## 19.1 Introduction

Intelligence is a controversial topic. It is the ability to process complex ideas, reason, and solve problems. Researchers and scientists have identified interpersonal and intrapersonal intelligences, which vary from person to person and from one culture to another. There is also musical/rhythmic intelligence [1]. Like other intelligences, musical intelligence can be developed through schooling and learning. What do Michael Jackson, Fela Ransome-Kuti, and Barbara Streisand all have in common? These individuals are considered to have strong musical intelligence; they think in music and rhythms.

Music touches our heart; it makes one happy, sad, upbeat, or relaxed. Friedrich Nietzsche once said, “Without music, life would be a mistake.” Music is an intellectual competence on its own. Its function can be found in a particular area of the brain. Figure 19.1 illustrates different types of music [2]. As a powerful learning tool, it is important for all of us to incorporate the art of music into our lives. People differ in the rate at which they acquire complex skills. Music is no exception.


M. N. O. Sadiku, S. M. Musa, *A Primer on Multiple Intelligences*,
https://doi.org/10.1007/978-3-030-77584-1_19

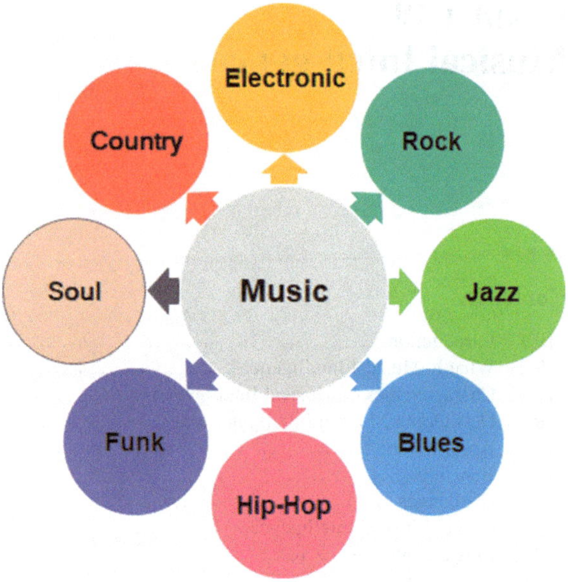


**Fig. 19.1** Different types of music

This chapter provides an introduction to musical intelligence (MI). It begins by describing what music intelligence is all about. It discusses some characteristics of musical intelligence and how someone can develop it. It presents some applications, benefits, and challenges of musical intelligence. It addresses MI in global context. The last section concludes with comments.

## 19.2 What Is Musical Intelligence?

Music pervades every facet of human experience. Like language, music is natural to humankind and is universal. Musical intelligence (aka music smart or sound smart) refers to the ability to perform, compose, and appreciate music. It may also be regarded as the "simulation of musical thinking." It deals with the ability to create or interpret music. Individuals with high musical intelligence have a strong appreciation of music and rhythm.

Musical rhythmic intelligence refers to how skillful an individual is performing, composing, and appreciating music. It is the ability to enjoy, observe, differentiate, create, form, and express musical forms. Musical intelligence is one of Howard Gardner's nine multiple intelligences [3].

- Naturalist (nature smart)
- Musical (sound smart)
- Logical-mathematical (number/reasoning smart)
- Existential (life smart)
- Interpersonal (people smart)

- Bodily-kinesthetic (body smart)
- Linguistic (word smart)
- Intrapersonal (self smart)
- Spatial (picture smart)

All of these intelligences can be developed to a high level. According to Gardner, intelligences is a way of knowing and understanding yourself and the world around you. He states that the musical intelligence emerges earlier than any of the other intelligences.

Musical intelligence is not the same thing as music talent. It consists of sensitivity towards rhythm, melody, and musical instruments. It is well known that the human brain has two halves, the left and the right hemispheres. Musical abilities are located mostly in the right hemisphere of the brain, but trained musicians are likely to draw upon the left hemisphere. People with high musical intelligence are gifted at composing, singing, or playing an instrument [4]. They find themselves being musician, singer, choir director, band director, composer, conductor, music teacher, pianist, vocalist, record producer, singer, songwriter, worship minister, disc jockeys, recording engineers, performers, dancers, musical instrument makers, rap artists, songwriters, record producers, music retailers, and music critics. Figure 19.2 shows an example of one playing an instrument [2].

Musical intelligence is related to emotional intelligence. By nature, musicians are of the emotional type. Their task is basically to play upon feeling, to appreciate, and to create the beautiful in the emotional realm. To be successful, the musician

**Fig. 19.2** A typical instrumentalist [2]

must carry his audience on a wave of emotion often bordering on the point of ecstasy. He specializes in his emotional control of the audience [5].

## 19.3 Characteristics of Musical Intelligence

One does not need to be a skilled musician to possess musical intelligence. Individuals with musical intelligence are highly auditory and very sensitive to rhythm. They discern sounds and tones with ease. They can hear and recognize patterns easily. They have a "good ear" for music and can easily learn songs. They think in terms of patterns and seek for patterns in speech and language. Some are especially gifted at playing instruments. Thus, people with musical intelligence have the following [6]:

- Seek patterns in their environment
- Be drawn to sound
- Easily memorize phrases and words in foreign languages
- Enjoy dancing and singing
- Use patterning to remember things
- Have good rhythm
- Be skilled at playing several instruments
- Be zealous about music
- Have the ability to easily remember songs
- Have a high level of understanding of musical structure, notes, tone, and rhythm
- Often listening to songs while studying
- Highly passionate about music
- Inclined towards learning and playing one or more musical instruments
- Uncanny talent to recognize and remember songs
- Create a musical autobiography by collecting recordings of music
- Love performing

In addition, individuals with high musical intelligence possess some or all of the following functional skills [7]: (1) vocal skills, (2) composing skills, (3) instrumental skills, and (4) listening skills.

## 19.4 Developing Musical Intelligence

How do you develop musical intelligence? What you need is an educational and family environment that fosters its development. The more you expose yourself to different sounds, tones, and rhythms, the stronger your musical intelligence will grow. Here are some ways you can grow in musical intelligence [8]:

- Play the "name that tune" game.

- Go to live music events
- Pick up a new instrument
- Add to your music collection
- Expose yourself to new genres and styles
- Listen to nature soundscapes

This list is by no means exhaustive.

Development of tests of musical intelligence and ability is based on the field of psychology, just like the Intelligence Quotient (IQ). Taking a musical intelligence test provides some benefits. Students can find out if a career in music would benefit them. Students may also find out why they are musically inclined. Musical intelligence is assessed in undergraduate and graduate programs. Before any test of musical aptitude is recommended, however, music educators must be convinced of its merits and must be well informed about the factors which can affect an examinee's scores.

## 19.5 Applications

The role of musical intelligence includes perceptions about music integration, musical growth, assessment of musical growth, and assessment in general. Decision-makers often downplay the importance of music in the curriculum. The use and development of musical intelligence should be encouraged in all areas and levels of education [9]. The goal of music education is not to produce a "musician," but to educate a human being.

Parents and teachers should have exposure to a variety of music and music listening as important experiences for developing musical intelligence in children. Figure 19.3 shows children's interest in music to increase the musical intelligence [10]. We now consider three areas in which musical intelligence is being applied.

- *Teaching:* Musical intelligence can be applied in the field of education and its possible benefits for students. Including music activities in a classroom can enhance the overall learning environment. Students can be encouraged to enhance their musical intelligence using the arts (music, art, theatre, dance). Figure 19.4 illustrates an example of musical intelligence teaching. As a teacher, you can enhance the musical intelligence of your students by doing the following [11]:

□ Including music in lessons where appropriate.
□ Playing background music for various activities in the classroom.
□ Allowing them to include music for independent projects.
□ Connecting music to a lesson, such as talking about what music was popular during historical periods.
□ Using songs to help students study for exams.
□ Using computer music software and games.
□ Playing Mozart or Beethoven as students study in class.
□ Playing musical instruments.

**Fig. 19.3** Children are interested in music to increase the musical intelligence [9]

**Fig. 19.4** Teaching musical intelligence [10]

- □ Encouraging students to add music to plays.
- □ Creating musical mnemonics.
- *Parenting:* By learning more about both learning and play behaviors of your children, you will be better able to teach and parent in a way that celebrates and embraces your child's personality type. Help them to play different styles of

music and different musical instruments like piano, violin, or guitar. Children have the capacity to learn different things (such as languages and music) simultaneously. When children are involved in musical activities, there is an increase in musical awareness, enjoyment, skill, and confidence, more positive musical attitudes, and enhanced overall learning. When a child has a high musical intelligence, it helps build the child's self-esteem and inner strength. Parents can play a role in developing the musical aptitude of their children by doing the following [2]:

  - Playing songs often according to their age
  - Encouraging and teaching them to play a musical instrument
  - Introducing them to various musical instruments
  - Creating games that incorporate music
  - Bringing them to music performances
  - Giving positive reinforcement when they are playing a musical instrument or singing
  - Helping them create musical mnemonics when they learn things

- *Language Learning:* This is important as valid communication can help interactions between countries. Learning foreign languages is difficult. For example, there are natural barriers between the English and Chinese languages which affect teaching approaches. The teaching of English in Chinese classrooms is removed from real-world English learning environments. A musical intelligence e-Learning approach can benefit Chinese students learning the English language. By nature, musical intelligence has parallel structures with language [12]. Music has been discovered as a powerful aid to language learning. Music affects logic as well as language.

## 19.6 Benefits

Music accompanies our everyday activities, makes learning easier, and facilitates language learning. It touches our emotion and influences our mood. It is something that brings us the most pleasure, comparable to food and sex. It improves the attention span of children while fostering their creativity. Musical intelligence is central to human experience. It plays a vital role in whole brain learning. Musical intelligence activities provide students with the chance to learn, practice, and apply music. Students can use musical intelligence activities to enhance their understanding of tone, rhythm, melodies, and timbre.

Studies have shown that listening to music greatly benefits the brain, sleep patterns, and stress levels in people. Music relaxes and helps man's meditation. Music tastes and personality are strongly related. In other words, your personality is based on the music you listen to. Musical intelligence helps a person find a pattern in everything they undertake. People with high musical intelligence can readily use rhythms and patterns to assist in learning. They have a sensitive ear for sound. It has been claimed that playing a musical instrument can make one smarter.

## 19.7 Challenges

Musical intelligence presents a number of modern problems. Intelligence is an abstract concept that creates much controversy. Defining intelligence is difficult and many researchers create a conflict in their definition of intelligence. There is criticism from people who defend a single factor of intelligence. Some critics argue that musical intelligence should be regarded as talent, not an intelligence. Music intelligence has not been widely studied as a personality variable in spite of its importance.

Implementing musical intelligence into elementary school curricula has been controversial. Political climate at the school and school district sometimes determine the role of musical intelligence in the school curriculum. Without an ongoing assessment of musical intelligence, there is no way of knowing whether music activities are helpful, harmful, or ineffective for elementary schools [13].

## 19.8 Global Musical Intelligence

The influence of music on virtually every aspect of life is complex. Music is universal and expressive, crossing cultural borders, playing a significant, unifying role in our modern society. Here we consider musical intelligence in different nations.

- *United States*: Music educators are concerned that music has often been marginalized in the public school curriculum of America. There is a huge disconnect between music practitioners and the real world. Music activities may be helpful, harmful, or ineffective for musical growth in schools. Educators, administrators, and parents have questioned traditional assessments of intelligence in America's schools. Standardized tests in the public schools have historically been concerned with only two categories of aptitude: verbal and mathematical. The implementation of musical intelligence into elementary curricula means many different things to different people [13].
- *Africa:* In an African musical event, everyone present plays a part and making music is never simply a matter of creating fresh improvisations, but a matter of expressing the sense of an occasion. Compared with Western music, African music has been epitomized as fundamentally a collective art. Ethnic identity and respect of Zimbabwean people is concerned with music. In South Africa, the government is ensuring that education is embarking on a course of major transformation, affecting educational structures as well as information presented to

learners, and music educators cannot afford to remain professionally uninvolved in this "historic revolution" [14].

- *Thailand:* In the past, most school curriculum around the globe considered IQ scores as the most important factor for a student's academic success. Today, many international schools in Thailand incorporate subjects such as music, drama, physical education, arts, and many more within the school curriculum to enhance learning in many domains. Musical intelligence can improve the academic performance, morale, and confidence of students. The teacher should help students discover their talents and help them grow [15].
- *Scotland:* Musical talent and musical intelligence can be developed within the existing educational infrastructure in Scotland. Music has both extrinsic and intrinsic value in the school curriculum. It has been observed that musical intelligence is highly developed in some students than in others due to genetic or environmental factors. Environmental factors might include opportunities provided through families, schools, and churches for performing, inventing, and listening to music [16].

## 19.9 Musical Intelligence Test

Musical intelligence reveals itself in many ways and it is developed over a time. The musical intelligence test is often used to identify a student's ability to enjoy, observe, differentiate, create, form, and express music. Taking a musical intelligence test provides several benefits. Students can find out if a career in music would benefit them. Students may also find out why they are musically inclined. Music teachers may use the test to determine the innate musical capabilities of their students.

The following test can be taken on range on a five-point scale [17]:

1. On occasions, I have felt almost moved to tears while listening to music.
2. I have often experienced euphoric feelings while listening to music.
3. I get upset if someone disturbs me while I am listening to music.
4. I tend to stop and notice sounds like an elevator chime or a ring tone.
5. I tend to stop and notice any rhythmic sound pattern such as the whirr of a printer.
6. I make efforts to seek out good music.
7. I make extra effort to attain sophisticated musical taste.
8. I can easily differentiate between major and minor keys. I can easily tell the difference between original version and an adaptation.
9. I can recall the original music even if I hear a version on another instrument.
10. I can predict the next sequence in an unfamiliar music, after hearing the first part.
11. I can easily recognize the logical structure underlying a musical composition merely by listening to it.

## 19.10 Conclusion

Musical intelligence consists of an individual's sensitivity to melody, harmony, rhythm, timbre, and musical structure. It is an ability or intelligence that we should be encouraged from a very early age. It involves emotional development that promotes empathy and expressing feelings. It is characterized by people who have a general musical bent of mind. Individuals with high musical intelligence have an ear for music. For more information about musical intelligence, one should consult the books in [18–21] and the following related journals:

- *Intelligence*
- *Journal of Music Theory*
- *Music Educators Journal*

## References


1. Musical/Rhythmic intelligence, http://www.edu-nova.com/articles/musical-intelligence/#:~:text=Researchers%20define%20musical%2Frhythmic%20intelligence,melodies%20and%20understand%20musical%20structure
2. Musical intelligence: characteristics, examples, and improvement, https://melodyful.com/musical-intelligence-characteristics-examples
3. H.E. Gardner, *Frames of Mind: The Theory of Multiple Intelligence* (Basic Book, New York, 1983)
4. M.N.O. Sadiku, S.M. Musa, A. Ajayi-Majebi, Essence of musical intelligence. Int. J. Trend Res. Develop. **7**(4) (2020)
5. C.E. Seashore, The psychology of music. XIII. Musical intelligence. Music. Educ. J. **24**(5), 32–33 (1938)
6. Musical intelligence: definition, experiments and characteristics, https://study.com/academy/lesson/musical-intelligence-definition-experiments-characteristics.html#:~:text=People%20with%20musical%20intelligence%20are,sensitive%20to%20rhythm%20and%20sound.&text=For%20example%2C%20they%20look%20for,patterns%20in%20speech%20and%20language
7. Multiple intelligences, http://www.inspiring-breakthrough.co.uk/learning-styles/intrapersonal-learning.htm
8. S. Terrell, Is musical intelligence your secret strength? Here's how to find out (2018), https://blog.mindvalley.com/musical-intelligence/
9. S. Snyder, Developing musical intelligence: why and how. Early Childhood Educ. J. **24**(3), 165–171 (1997)
10. Musical Monday', an event that enables young people to make music together, returns later this month (2017), https://www.mytownsouthend.com/single-post/2017/10/19/%E2%80%98Musical-Monday%E2%80%99-an-event-that-enables-young-people-to-make-music-together-returns-later-this-month
11. Teaching students who have musical intelligence, https://www.boldsky.com/pregnancy-parenting/kids/2014/how-to-enhance-musical-intelligence-in-children-056047.html
12. L. Wu, M. McMahon, Adopting a musical intelligence and eLearning approach to improve the English language pronunciation of Chinese students. IFAC Proc. **45**(10), 160–164 (2012)
13. S.W. Mills, The role of musical intelligence in a multiple intelligences focused elementary school. Int. J. Educ. Arts **2**(4) (2001)

14. J. Matare, Creativity or musical intelligence? A comparative study of improvisation/improvisation performance by European and African musicians. Think. Skills Creat. **4**, 194–203 (2009)
15. L. Sachathep, R. Lynch, A comparative - correlational study of emotional intelligence and musical intelligence among students from years eight to eleven at modern international school Bangkok, Thailand. Scholar **8**(1), 53–62 (2016)
16. P. Stollery, A.D. McPhee, Some perspectives on musical gift and musical intelligence. Br. J. Music Educ. **19**(1), 89–102 (2002)
17. V. Krishnan et al., Musical intelligence: explication, measurement, and implications for consumer behavior. J. Consum. Mark. **31**(4), 278–289 (2014)
18. L. Montello, *Essential Musical Intelligence: Using Music as Your Path to Healing, Creativity, and Radiant Wholeness* (Quest Books, 2002)
19. J. Bamberger, *The Mind Behind the Musical Ear: How Children Develop Musical Intelligence* (Harvard University Press, Cambridge, MA, 1991)
20. A. Bowe, *Musical Intelligence as Creative Flow* (Lulu, 2013)
21. E. Liebman, *Sequential Decision-Making in Musical Intelligence* (Springer, 2019)

# Chapter 20
# Military Intelligence

## Contents



> Successful organizations, including the Military, have learned that the higher the risk, the more necessary it is to engage everyone's commitment and intelligence.
>
> —Margaret J. Wheatley

## 20.1 Introduction

Modern warfare requires an integration of military and intelligence forces. Military intelligence is crucial and central to planning a victorious campaign. Before the commander could determine how to employ his forces, he first has to know whether he can attack and where he should attack. Military intelligence uses information collection to provide guidance and assist commanders in their decisions.

A battle is a physical activity that requires force. The victor could only employ the forces necessary to achieve victory through the advantage of foreknowledge. Winning a war also requires foresight, analysis, and the development of strategies on how to win—it takes intelligence. From time immemorial, intelligence has been a key factor in security planning. Intelligence has received increasing attention over the last decade [1]. While the policing is civilian (non-military) and focuses their intelligence work on criminals who posed threats to local communities, military intelligence focuses on military threats to the national security.



M. N. O. Sadiku, S. M. Musa, *A Primer on Multiple Intelligences*,
https://doi.org/10.1007/978-3-030-77584-1_20

This chapter provides a primer on military intelligence. It begins by discussing what military intelligence is all about. It describes the characteristics and principles inherent in military intelligence. It provides levels and sources of military intelligence. It discusses the necessity of military intelligence analysis. It gives some applications of military intelligence. It highlights the benefits and challenges of military intelligence. The last section concludes with comments.

## 20.2 What Is Military Intelligence?

Military intelligence (MI) is a military branch that uses information collection and analysis approaches to provide guidance and direction to assist commanders in their decisions. As an academic field, military intelligence is multidisciplinary area that combines language, political theory, economics, sociology, and psychology. Military intelligence includes information on other nations' military forces, plans, and operations which is collected through a variety of means. The term "intelligence" is often used for an agency that gathers such information. In the military jargon, intelligence is the art of "knowing one's enemies." The primary objective of military intelligence is to provide timely, relevant, and accurate intelligence support to tactical, operational, and strategic needs of the army.

Military/naval intelligence is as old as warfare itself. It is vital for the outcome of battles. In biblical times, Moses sent out 12 men to spy the Canaanites in order to learn about their strategies, attitudes, strengths, and weaknesses. The development of modern intelligence agencies dates back to the end of the nineteenth century.

The US Department of Defense (DoD) was created in 1949. It comprises the Department of Army, Office of the Secretary of Defense, the Joint Chiefs of Staff, the military departments (Army, Navy, and Air Force, each under the authority of a civilian Secretary), 16 "defense agencies" which were created by the Secretary of Defense to perform particular functions, nine "Unified Commands" responsible for the conduct of military operations, civilians, and contractors [2].

Today, nations have at their disposal information gathering systems such as radio, TV, satellites, ultramodern aircraft, human sources, cameras, and electronic devices. The United States government has devoted enormous resources to the creation and maintenance of a national intelligence system. The national intelligence communication includes the National Security Agency (NSA), Central Intelligence Agency (CIA), National Intelligence Strategy (NIS), Federal Bureau of Investigation (FBI), Defense Intelligence Agency (DIA), elements within the Department of Defense. The FBI is responsible for countering the activities of foreign intelligence services within the United States. The CIA is responsible for coordinating US counterintelligence activities abroad. The DIA produces some human intelligence and measurement and signatures intelligence. The NSA produces signals intelligence.

As mentioned earlier, US intelligence has two broad functions: collection and analysis. Using various means, intelligence agencies collect information about foreign nations, persons, places, events, and activities. Analysis consists of assessment

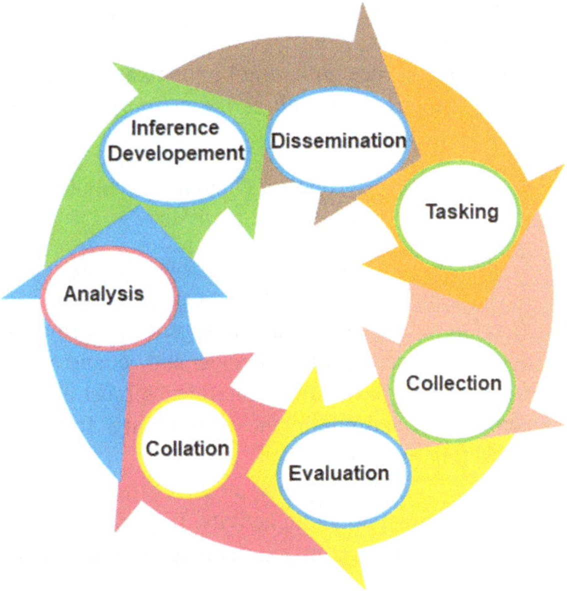


**Fig. 20.1** The intelligence cycle

of an adversary's capabilities and vulnerabilities. It helps an analyst to know the capabilities of common types of military units [3]. Collecting information in itself does not result in obtaining intelligence. Information must be properly evaluated before it can be acted upon. Thus [4]:

$$\text{INFORMATION} + \text{EVALUATION} = \text{INTELLIGENCE}$$

Intelligence involve the collection, evaluation, analysis, integration, collation, inference development, and dissemination of all available information which concerns areas of planning and operations. The intelligence cycle is usually presented graphically as a circular process, as shown in Fig. 20.1. Military intelligence systems can be classified according to the nature of their location (such as ground or space) or the nature of their mission (e.g., operational, tactical, or strategic).

## 20.3 MI Characteristics and Principles

The main characteristics of military intelligence are presented as follows [5]:

- To be effective, military intelligence has to be integrated with weapon systems and personnel to create military capability.
- Intelligence consists of collecting and assessing knowledge which is relevant to all levels of the military and the civilian leaders.
- Gaining information superiority is based on high-quality human capital and technology. • The intelligence mission is hampered by an inherent uncertainty, which

may be reduced by good procedures and high-quality human capital, but cannot be eliminated.
- Good intelligence gathering by the rival can reduce the relevance of the nation's own intelligence as well as the potency of its weapon systems and its overall military capability.

Not bounded by time or changing technology, the principles of intelligence include [6]:

1. *Accuracy*: With accurate intelligence on the enemy's location, capability, and intent, planning can lead to victory.
2. *Timeliness*: Timeliness of intelligence contributes directly to the commander's ability to observe, orient, decide, and act and achieve victory.
3. *Usability*: To have value, intelligence must be tailored to meet the needs of the commander and intelligence community. To make intelligence usable, effective marketing must be applied.
4. *Fusion*: Fused intelligence produces the most accurate and complete picture of what is known about an activity. It draws upon the complementary strengths of signals intelligence, imagery intelligence, human intelligence (all to be explained shortly).
5. *Relevancy:* Intelligence has relevancy if it contributes to the commander's ability to execute his mission at his level of military operation. It must be geared to meet the needs of commanders across the full spectrum of military operations.
6. *Intellectual honesty:* Intellectual honesty must extend to reporting. Intellectual honesty must drive the intelligence professional to distinguish for the commander information which is solidly grounded in fact. The commander cannot be left with uncertainty in his mind regarding what is fact and what is opinion.
7. *Communication*: It is the responsibility of the intelligence professional to employ the most effective way to inform the commander. Knowing the enemy's intent, capabilities, and location has no worth if the intelligence is not well communicated.

These seven principles form the basics of the art of military operations. They can greatly enhance the conduct of military operations at the strategic, operational, and tactical levels.

## 20.4 Levels of Military Intelligence

Military intelligence activities take place at all levels, from tactical to strategic, in peacetime and during a war.

- *Strategic intelligence*: This is the intelligence required for the formation of policy and military plans at national and international levels.

- *Operational intelligence:* This is intelligence required for planning and conducting campaigns and major operations. The operational tier is below the strategic level of leadership.
- *Tactical Intelligence*: This is the intelligence required for the planning and conduct of tactical operations. It is needed by military field commanders to plan and conduct combat operations if necessary. It is operational and tactical intelligence, not necessarily numbers or employment of force, that can achieve success in wartime operations.
- *Human intelligence:* This form of intelligence, gathered by spies, is often the only form of intelligence that provides information about an opponent's intentions. When providing on-site intelligence, planners were aided by human intelligence.

## 20.5 Sources of Intelligence

It is customary for the intelligence services of a nation to read the main newspapers and journals of the nations is. This is a basic source of intelligence. It is also common for diplomats and journalists personnel to have a secondary goal of collecting military intelligence for their nation. Some nations eavesdrop continuously on the entire radio spectrum, TV broadcasts, military traffic, radar emissions, telephone, and satellite traffic [7]. Different factors can be important in military intelligence. These factors lead to the following sources of intelligence as shown in Fig. 20.2 and explained as follows [8].

- Communications intelligence is derived from foreign communications that are intercepted by non-intended recipients.
- Electronics intelligence is technical information obtained from foreign electromagnetic radiation.
- Telemetry intelligence is technical information that is derived from intercepting foreign telemetry data.
- Human intelligence is provided by spies, who usually possess information about a nation's political leaders, strategies, and decisions.
- Biographical intelligence is information collected on the views, traits, habits, and story about the leaders and important individuals of a nation.
- Cartographic intelligence is derived from maps and charts.
- Economic intelligence is information concerning the production, distribution, and consumption of goods and services.
- Energy intelligence addresses the location and size of foreign energy resources.
- Counterintelligence is intended to detect, counteract sabotage, terrorist attacks, or assassinations.
- Geographic intelligence is derived from studying geographic characteristics such as terrain, climate, natural resources, transportation, boundaries, and population distribution.

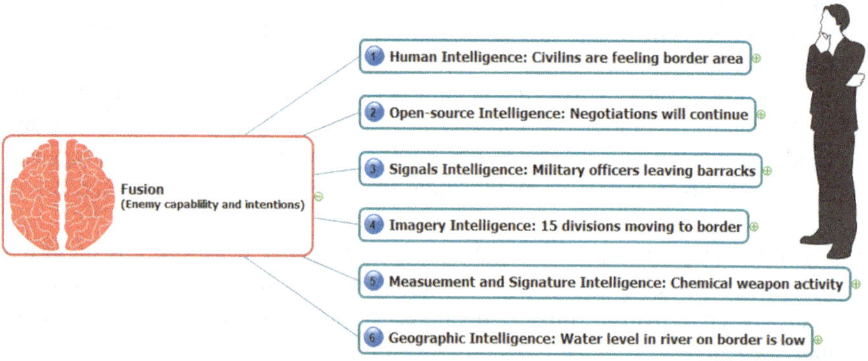


**Fig. 20.2** Factors that lead to sources of intelligence

- Medical intelligence is obtained from studying foreign natural and man-made environments that could affect the health of military forces.
- Sociological intelligence is gained from studying a nation's social stratification, value systems, beliefs, and other social characteristics.
- Transportation intelligence involves assessing a nation's ability to wage war, concerning its highways, railroads, inland waterways, and civil airways.

## 20.6 Military Intelligence Analysis

The main functions of military intelligence include collection, analysis, and dissemination of military information. Intelligence analysts use information collected and produce "all source" analysis for the commander. They generally use an intuitive approach to analysis and not structured techniques. They support military operations by providing critical intelligence, finished intelligence products, and crucial information to the commanders. The commander uses the intelligence for planning before deployment, while en route, and during operations. Intelligence must not only keep pace with threats but stay one step ahead and forecast them. This forces intelligence organizations to continuously update their structures, procedures, doctrines, and operations [9].

Producing military intelligence analysis is the responsibility of the Defense Intelligence Agency (DIA). Military intelligence analysts serve to meet the contemporary needs of commanders. They need high-quality intelligence to accomplish their missions. They do not have the luxury of detachment from practice and action. Military intelligence analysis has been scrutinized for its accuracy and value. In some countries, it has become outdated and needs revision. The analysis can use online information that is available free of charge from the Internet. The Global Positioning System (GPS) adds an inexpensive, useful tool needed to pinpoint enemy threats. Having the technology, personnel, and skills do not necessarily mean that commanders would act accordingly. According to some intelligence analysts, future wars will be "hybrid wars."

## 20.7 Applications

The following are common applications of military intelligence.

- *Finding Your Enemy*: Military intelligence has to do with knowing your enemies. As long as humans go to war to settle conflict, there will be a need for decision support to military and civilian leaders regarding enemies. Enemies will continue to gain influence, taking advantage of changing conditions in the international environment. The commander needs intelligence about the enemy and the battlespace prior to engaging in operations. Espionages aim at the goal of discovering the enemy's numbers, location, and formations. Secrecy is everything to them. It is also common for diplomatic and journalistic personnel to have a secondary assignment of collecting military intelligence.
- *General Military Intelligence* (GMI): GMI focuses on the military capabilities of foreign nations and organizations. It deals with issues relating to Armed Forces capabilities, including threat characteristics and terrain intelligence. The GMI database supports the unit's plan, preparation, execution, and assessment of operations [10].
- *Homeland Security*: Military intelligence services will support the military's use of force at home in case of civil disturbance or terrorist attack on the US homeland. What we do not know is whether an aggressive use of military intelligence at home contributes to current counterterrorism efforts of the FBI and local law enforcement. For example, federal troops were deployed in the 1950s and 1960s to help integrate Southern schools and deal with civil disorders in Detroit in 1967 [11]. Since September 11, 2001, terrorist attacks on the American homeland have changed things substantially.
- *Intelligence Organizations:* There are a great number of intelligence organizations at the national and international levels, making significant contributions by providing support to current operations. Military intelligence analysis is usually provided by the US Defense Intelligence Agency (DIA), created in 1983. Other intelligence organizations include the National Security Agency, the Central Imagery Office, and the Central Intelligence Agency. These organizations provide specialized support and advise defense policymakers on political-military matters [1].
- *Military Intelligence Specialists:* To become a military intelligence specialist, one must first join the armed forces. Military intelligence specialists are also required to have college degrees, strong communication and decision-making skills. Their main task is to gather, research, and analyze the information or data necessary for successful military operations. The data helps military leaders determine the most advantageous course of action. Military intelligence special-

ists can be assigned to various locations and their job includes providing leadership, prioritizing activities, and making presentations and reports to senior staff. They are sometimes trained as interrogators, interpreters, translators, radio announcers, and propaganda writers [12]. Figure 20.3 illustrates a typical military intelligence training [13].

## 20.8 Benefits

Military intelligence is the commander's decision-making tool. It helps the commander to make decisive predictions on when and where an action will take place. It helps to identify guerrilla combatants as well as "enemy" civilians. It can help lead to victory because it provides information on the adversary's capabilities and vulnerabilities. It reduces the unknowns that planners must face. It provides essential information that often saves the soldiers. If the military commander knows the strength of his opponent and the route of attack, he has the ability to consolidate and position his forces on the most advantageous ground. Yet force alone cannot win the day. Military intelligence gives the commander the knowledge of the battlefield and the understanding of their foe so that they can focus their forces at the right place to win [14].

Military intelligence also plays an important role in a knowledge race as well as arms race between two nations. Intelligence superiority produces an advantage in deterrence and preemption. One of its objectives of military intelligence is to warn of potential or immediate threats. It aims at insuring national security during times of peace and war. The US military intelligence community provides tactical

**Fig. 20.3** Military intelligence training [13]

intelligence, personnel, and resources to UN ground commanders. Thus, military intelligence is not a dispensable luxury but an indispensable necessity in modern military operations.

## 20.9 Challenges

Military intelligence gathering remains a controversial issue in international law and invites hot debates. Human intelligence, due to spying, is usually prone to inaccuracy. When a spy is hired by a country, the spy may defect for economic reasons. Battle is always risky. The intelligence field usually involves risking human lives. Illegal surveilling of journalists and government critics has been commonplace in Colombia for at least 15 years. Sometimes, military intelligence fails to provide commanders and policymakers with a good understanding of complex counterinsurgency environments. The bureaucratic nature of the intelligence community needs reformation.

Since intelligence operations are conducted at all levels in peacetime and during a war, intelligence is an expensive task. The rapid development of military intelligence is attended by rapidly increasing expenditure on intelligence operations, innovation, and equipment. For example, the US intelligence community is believed to be the world's largest, spending over $80 billion in 2010 [15].

## 20.10 Global Military Intelligence

Most national governments have a military intelligence capability to collect information in both specialist units and from other arms and services. Each country has to produce its own military R&D and the intelligence apparatus. The North Atlantic Treaty Organization (NATO) is an intergovernmental military alliance between 30 North American and European countries. NATO is a security alliance that fields military forces able to operate together in any environment. It will engage to prevent crises and manage crises where possible and when necessary. Current NATO military intelligence doctrine is fundamentally unchanged, training is unsuited to the operational environment now faced, and the organizational culture still pays limited attention to mission effectiveness.

We will typically consider the following countries and their military intelligence.

- *US Intelligence:* The US intelligence community is known as the world's largest, with an annual budget of $75 billion. An extensive intelligence system has been a vital part of the United States government since 1941. The primary mission of national military intelligence in the United States Army is to provide timely, accurate, and synchronized electronic warfare support to tactical, operational, and strategic-level commanders. In 1942, the Military Intelligence Division was reorganized as the Military Intelligence Service. It was tasked with collecting, analyzing, and disseminating intelligence. The great military victory we achieved in Desert Storm and the minimal losses sustained by US and Coalition forces can be directly attributed to the excellent intelligence picture we had on the Iraqis. In 1962, the Army Intelligence and Security Branch was established as a basic Army branch to meet the increased need for national and tactical intelligence [16]. The USA emerged from the Cold War as the world's only multidimensional (e.g., military, economic, and political) "superpower." The wars fought by the United States have become asymmetric or irregular in nature, i.e., an armed struggle between combatants that does not fit the conventional idea of warfare. Figure 20.4 shows US Marines with Second Intelligence Battalion [17].

- *Israeli Intelligence:* Israel is the only liberal democracy today where a military intelligence service leads in both military and civilian affairs. Israeli intelligence was outstanding in the Six-Day War in June 1967. Israel could not count on technological superiority to overcome the odds. But Israeli intelligence was superb, having located every Egyptian squadron, revealed the layout of every air base, and mastered Egyptian Air Force operational procedure. Using this information, Israeli intelligence developed a precise targeting scheme. This shows how strategic intelligence can be used in conjunction with operational intelligence to provide senior decision-makers information. Yet, just as Israeli intelligence in this case can be viewed as an example of how intelligence operations should be conducted, Cairo's intelligence services opened the door to its own defeat. While Israeli forces were better trained, had superior leadership, and had a far more flexible doctrine, Egypt's army could boast that the majority of its soldiers were combat veterans [18]. The Israeli Directorate of Military Intelligence, AMAN, is not merely focused on collection and research regarding military intelligence

**Fig. 20.4** US Marines with Second Intelligence Battalion March in 2015 [17]

matters. Instead, AMAN covers the majority of intelligence activity arenas, including intelligence regarding state-related issues [19].

- *Canadian Intelligence:* The evolution of a military intelligence system in Canada developed largely due to the influence of the two world wars. Rather than consolidating the wartime intelligence community to serve postwar needs, Canada submitted itself to subordinate partnership in an intelligence alliance with a number of countries such as the United States, the United Kingdom, Australia, and New Zealand [20]. The Canadian Forces (CF) face an uncertain future due to some factors such as its limited size and resources and operational "overstretch." There are few qualified intelligence specialists and they are spread far too thin [20].
- *Chinese Intelligence*: China's civilized history dates back nearly five millennia. However, the People's Liberation Army (PLA) has existed for less than a century. The PLA has been a politicized "party army" from the very beginning. China enjoys growing international standing and influence. Its leaders see China's long-term military modernization program as essential to achieving great power status. China's comprehensive national strength, core competitiveness, and risk-resistance capacity are increasing. China realizes that the PLA must be prepared to fight in new realms and adapt to the modern, high-tech battlefield. This realization permeates China's national security strategy. The PLA engages with foreign militaries to demonstrate its growing capabilities and enhance China's image and influence abroad. China's armed forces intend to develop relations with their counterparts in neighboring nations and continue the traditional friendly military ties with their African, Latin American, and Southern Pacific counterparts [21]. Figure 20.5 depicts Chinese military intelligence [22].
- *British Intelligence:* The Intelligence Corps is one of the smallest Corps in the British Army. It covers the entire range of potential intelligence from strategic government level to operational contingency planning. It is responsible for gathering, analyzing, and disseminating military intelligence. Operator Military Intelligence soldiers are employed in a variety of roles within the British Army [23].
- *Nigerian Intelligence:* Nigeria is an African country with a land mass of about 923,768 sq. kms. Since 1999, following the end of several years of military rule, Nigeria has been confronted with the problem of national security arising from violent crimes, including armed robbery, assassinations, ethno-religious violence, kidnapping for ransom, and terrorist bomb blasts. Boko Haram terrorists have conducted about 136 attacks, largely on police stations, churches, and mosques. The Boko Haram terrorism has negatively affected national security, democracy, good governance, and economic development [24].
- *Indian Intelligence*: The Defence Intelligence Agency, created in March 2002, is an organization responsible for providing and coordinating military intelligence for the Indian armed forces. It operates within the Ministry of Defence. The agency has an ongoing and expanding bilateral cooperation with the United States military. In order to upgrade the obsolete systems, the agency floated the process of acquiring new photogrammetry and geographical information sys-

**Fig. 20.5** Chinese military intelligence [22]

tems. These equipment were essential for getting better clarity from satellite images to track the movements across the border, including military exercises, training, and military buildup by China and Pakistan [25].

- *Syrian Intelligence:* The General Intelligence Directorate is the major civilian intelligence service in Syria. It is headquartered at the Defense Ministry in Damascus. It consists of three powerful divisions: internal (equivalent to the FBI), external (equivalent to the CIA), and Palestine, charged with monitoring Palestinian activities. Although its primary role is quelling internal dissent, it is often suspected for providing support to different radical groups [26].

## 20.11 Conclusion

Looking towards the future, one can imagine a fundamental change in the character of war. Military intelligence will continue to play a decisive role in future battles. It should not be relegated to a secondary status. The defense intelligence community will require many enabling technologies from private industry to help it meet its goals.

Job training for a military intelligence officer requires completing some courses. American Military University, which was founded in 1991, provides military personnel with portable, relevant, and affordable education. Qualified students can earn full-tuition, merit-based scholarships, and stipend for living expenses.

More information about military intelligence can be obtained from the books in [27–51] and other numerous books available at Amazon.com and books.google.com. One should also consult the journals that are exclusively devoted to it:

- *Studies in Intelligence*
- *Military Intelligence Digest*
- *Canadian Military Journal*
- *The Military Engineer*
- *American Intelligence Journal*
- *Intelligence*

## References


1. M.N.O. Sadiku, O.D. Olaleye, A. Ajayi-Majebi, S.M. Musa, Military intelligence: a primer. Int. J. Trend Res. Develop. **7**(3), 298–302 (2020)
2. Military intelligence, https://fas.org/irp/offdocs/int014.html
3. The role of intelligence, https://fas.org/irp/offdocs/int006.html
4. Criminal intelligence: manual for analysts, https://www.unodc.org/documents/organized-crime/Law-Enforcement/Criminal_Intelligence_for_Analysts.pdf
5. E. Pecht, A. Tishler, The value of military intelligence (2011), https://www.researchgate.net/publication/228423002_The_Value_of_Military_Intelligence
6. D.G. Payton, The art of intelligence by the General, https://apps.dtic.mil/dtic/tr/fulltext/u2/a283160.pdf
7. Military intelligence, Wikipedia, the free encyclopedia, https://en.wikipedia.org/wiki/Military_intelligence
8. B.W. Watson, Intelligence: military science, https://www.britannica.com/topic/intelligence-military
9. G.E. Lewis, *Army Intelligence Analysis Transforming Army Intelligence Analysis Training and Doctrine to Serve the Reasonable Expectations and Needs of Echelons Corps and Below Commanders, Consumers, and Customers* (United States Army, Leavenworth, Kansas, 2005)
10. TC 2-50.5: intelligence officer's handbook, https://www.coursehero.com/file/54569953/TC-2-505-Intelligence-Officers-Handbookpdf/
11. S. Dycus, The role of military intelligence in homeland security. Louis. Law Rev. **64**(4) (2004)
12. Become a military intelligence specialist: education and career roadmap (2020), https://study.com/articles/Become_a_Military_Intelligence_Specialist_Education_and_Career_Roadmap.html
13. M. Everly, Technology enhances military intelligence training, capabilities for soldiers (2017), https://www.army.mil/article/196669/technology_enhances_military_intelligence_training_capabilities_for_soldiers
14. G. Elder, Intelligence in war: it can be decisive: winning with intelligence, https://www.cia.gov/library/center-for-the-study-of-intelligence/csi-publications/csi-studies/studies/vol50no2/html_files/Intelligence_War_2.htm
15. E. Pecht, A. Tishler, The value of military intelligence. Def. Peace Econ. **26**(2), 179–211 (2015)
16. Military intelligence corps (United States Army), Wikipedia, the free encyclopedia, https://en.wikipedia.org/wiki/Military_Intelligence_Corps_(United_States_Army)
17. https://www.iimef.marines.mil/Photos/igphoto/2001316784/
18. E. Pascovich, Military intelligence and controversial political issues: the unique case of the Israeli military intelligence. Intell. Nat. Secur. **29**(2), 227–261 (2014)
19. W.K. Wark, The evolution of military intelligence in Canada. Armed Forces Soc. **16**(1) (1989)
20. D.A. Charters, The future of military intelligence within the Canadian forces. Canad. Milit. J., 47–52 (2001–2002)
21. Defense Intelligence Agency, China military power modernizing a force to fight and win (2020)
22. China fielding world's most advanced weapon systems: Pentagon report, https://www.news18.com/news/world/china-fielding-worlds-most-advanced-weapon-systems-pentagon-report-2003715.html

23. Intelligence corps (United Kingdom), Wikipedia, the free encyclopedia, https://en.wikipedia.org/wiki/Intelligence_Corps_(United_Kingdom)
24. F.E. Oghi, J.O. Unumen, Military intelligence and the challenge of national security in contemporary Nigeria. Int. J. Res. Human Soc. Stud. **1**(2), 7–13 (2014)
25. Defence Intelligence Agency (India), Wikipedia, the free encyclopedia, https://en.wikipedia.org/wiki/Defence_Intelligence_Agency_(India)
26. General Intelligence Directorate (Syria), Wikipedia, the free encyclopedia, https://en.wikipedia.org/wiki/General_Intelligence_Directorate_(Syria)
27. S. Tzu, *The Art of War* (Oxford University Press, New York, 1963)
28. F. Baudet et al., *Perspectives on Military Intelligence from the First World War to Mali: Between Learning and Law* (Springer, 2017)
29. S.M. Harris, *British Military Intelligence in the Crimean War, 1854–1856* (Routledge, New York, 2020)
30. P. Leverkuehn, *German Military Intelligence* (Pickle Partners Publishing, 2016)
31. D.R. Mallett, *Hitler's Generals in America: Nazi POWs and Allied Military Intelligence* (University of Kentucky Press, Lexington, 2013)
32. D.M. Glantz, *Soviet Military Intelligence in War* (Routledge, New York, 1990)
33. Y. Shefft, *British Military Intelligence in the Palestine Campaign 1914–1918* (Routledge, New York, 1998)
34. J.P. Finnegan, *Military Intelligence* (Center of Military History, United States Army, Washington DC, 1998)
35. P.A. Mohs, *Military Intelligence and the Arab Revolt: The First Modern Intelligence War* (Routledge, New York, 2010)
36. G. W. Hopple, B. W. Watson (eds.), *The Military Intelligence Community* (Routledge, New York, 2019)
37. J.M. House, *Military Intelligence, 1870–1991: A Research Guide* (Greenwood, 1993)
38. J.L. Gilbert, *World War I and the Origins of US Military Intelligence* (The Scarecrow Press, Plymouth, 2012)
39. J.P. Finnegan, *Military Intelligence: A Picture History* (U.S. Army Intelligence and Security, 1992)
40. E. Sergeev, *Russian Military Intelligence in the War with Japan, 1904-05: Secret Operations on Land and at Sea* (Routledge, London and New York, 2007)
41. J.T. Richelson, *A Century of Spies: Intelligence in the Twentieth Century* (Oxford University Press, 1997)
42. R.W. Leonard, *Secret Soldiers of the Revolution: Soviet Military Intelligence, 1918–1933* (Greenwood Press, Westport, CT, 1999)
43. P. Dimitrakis, *Military Intelligence in Cyprus: From the Great War to Middle East Crises* (I. B. Tauris, 2010)
44. J. Moran, *From Northern Ireland to Afghanistan: British Military Intelligence Operations, Ethics and Human Rights* (Routledge, New York, 2016)
45. T.M. Proctor, *Female Intelligence: Women and Espionage in the First World War* (New York University Press, New York, 2003)
46. J. Hughes-Wilson, *Military Intelligence Blunders* (Da Capo Press, 2000)
47. G.W. Hopple, B.W. Watson, *The Military Intelligence Community* (Routledge, New York, 2020)
48. H.V.F. Winstone, *The Illicit Adventure: The Story of Political and Military Intelligence in the Middle East from 1898 to 1926* (J. Cape, 1982)
49. E.C. Fishel, *The Secret War for the Union: The Untold Story of Military Intelligence in the Civil War* (Moughton Mifflin, New York, 1996)
50. J.A. McChristian, *The Role of Military Intelligence, 1965–1967* (Department of the Army, Washington DC, 1975)
51. *US Military Intelligence Handbook, Volume 1 Strategic Information, Procedures and Developments*, 6th edn. (IBP USA, 2016)

# Appendix: Other Intelligences

Academic Intelligence
Agriculture Intelligence
Animal Intelligence
Artistic Intelligence
Biblical Intelligence
Blended Intelligence
Brain Intelligence
Cognitive Intelligence
Collaborative Intelligence
Collective Intelligence
Community Intelligence
Competitive Intelligence
Componential Intelligence
Cosmic Intelligence
Crystallized Intelligence
Customer Intelligence
Cyber Intelligence
Decision Intelligence
Distributed Intelligence
Einstein Intelligence
Existential Intelligence
Financial Intelligence
Fluid Intelligence
Geospatial Intelligence
Global Intelligence
Governance Intelligence
Grey Intelligence
Hot Intelligence
Imaginational Intelligence


M. N. O. Sadiku, S. M. Musa, *A Primer on Multiple Intelligences*,
https://doi.org/10.1007/978-3-030-77584-1

Industrial Intelligence
Integrated Intelligence
Intellectual Intelligence
Intelligence Everything
Internet Intelligence
Mating Intelligence
Mental Intelligence
Mobile Intelligence
Moral Intelligence
Negative Intelligence
Network Intelligence
Pedagogical Intelligence
Personal Intelligence
Plant Intelligence
Practical Intelligence
Private Intelligence
Public Intelligence
Rational Intelligence
Relational Intelligence
Religious Intelligence
Sport Intelligence
Successful Intelligence
Verbal Intelligence
Visual Intelligence
Web Intelligence

# Index




M. N. O. Sadiku, S. M. Musa, *A Primer on Multiple Intelligences*,
https://doi.org/10.1007/978-3-030-77584-1